HEALESVILLE SANCTUARY

A FUTURE FOR AUSTRALIA'S WILDLIFE

HEALESVILLE SANCTUARY

A FUTURE FOR AUSTRALIA'S WILDLIFE

Sally Symonds

PIZZEYWALKER

ARCADIA

© Sally Symonds 1999

First published 1999 by
PizzeyWalker ARCADIA
Australian Scholarly Publishing Pty Ltd
P.O. Box 299, Kew, Victoria 3101
Tel: 03 – 9817 5208
Fax: 03 – 9817 6431

Printed by Imprenta Pty Ltd.

Acknowledgements

First, a special tribute: to those people so dedicated to the task in hand that they count the cost neither in time nor money. For these are the people to whom the Sanctuary owes a great debt for its very existence.

Robert Eadie and David Fleay were among many of that kind in the early years. Kevin Mason has kept the torch alight today, caring for the Sanctuary with the same deep commitment. To him I am indeed indebted for his unstinted efforts to preserve the past and his generosity in sharing his knowledge.

To Rosemary Fleay-Thomson I am grateful for access to the papers and photographs of David Fleay's years at the Sanctuary, and to the Fleay family and David Fleay Trustees for permission to publish his photographs. My thanks too to the family of Robert Eadie for sharing the pages of his memoir, and to Graham Endacott for access to the Morwell Hodges collection.

I would like to thank the staff at Healesville Sanctuary, past and present, for their time, their knowledge and their ideas. I am grateful too to Robert Fleay, Jean Osborne, Charles Stanley and the many other individuals who have contributed from their personal recollections; and to Richard Butler and Marilyn Warne-Smith for reading and advising on the manuscript.

My thanks to all the photographers who made their material available: from those of the early days whose work is in the Healesville Sanctuary collection, to those of more recent years. These include John Gollings, Mark Griffin, Gary Lewis, Peter Marsack, Kevin Mason, Steve Parish, David Scaletti, Michael Silver and Ian Smales, Len Smith and Bob Winters.

I appreciate the assistance given by the Friends of the Zoos, and John Alsemgeest at the Healesville Historical Society; and the support of the Zoological Parks and Gardens Board for the production of this book.

Sally Symonds
Richmond

CONTENTS

Robert and Rosemary Fleay with a family of emus

FOREWORD

In the years following the second world war, the Healesville Sanctuary was a favourite destination for country train excursions and Sunday driving. At the Sanctuary you could walk among kangaroos and emus in outdoor enclosures, surrounded by tall, white-trunked manna gums; look up at the dark, timbered bulk of Mt Toolebewong and smell the spicy peppermint of eucalypts.

There were pythons and spotted cuscuses from New Guinea, wombats, dingoes, exquisitely marked spotted harriers and drowsy powerful owls with glowing bright-yellow eyes; immaculate white goshawks and even a trained brown goshawk called Snapdragon. One wonderful day David Fleay, then the Sanctuary's director, even flew his trained peregrine falcon for me, a privilege I have never forgotten. I can still see that sickle-winged shape coming in to attack a swinging lure at eye-level, at speeds probably well over a hundred kilometres per hour.

The Sanctuary was magical, and now modernised, with sophisticated teaching and display facilities, it still has the capacity to capture the hearts and minds of visitors, as it did mine.

This history by Sally Symonds is a welcome reminder of the social and national value of a wonderful place.

GRAHAM PIZZEY, AM
DUNKELD

1

THE SETTING FOR A SANCTUARY

In the moonlight a ring of water spreads across Badger Creek as a platypus dives to the depths. The broad flat beak sways from side to side, sensing out a worm here, a yabby there, along the creek bed.

A shriek, and a shadow passes overhead, a yellow-bellied glider leaping from one tree to another, its outstretched body and fluffy tail etched against the moonlit sky. A rustle in the bushes and the portly shape of a wombat emerges and lumbers along the pathway.

At night, time has stood still along the Badger. A dingo's howl still rends the night air; the thump of a wallaby moving through the undergrowth, the snuffle of a bandicoot as it digs for tubers, small rustles of antechinus and bushrat. The mournful whoo-hoo of the powerful owl floats through the forest.

Animals have for centuries rested in the shade of the tall gums which line its banks, built their homes in the hollows of its old trees, or among the shrubs and tree ferns. Aborigines have hunted through the undergrowth, cut their canoes and shields from the bark of its trees and fished in its waters. Europeans too have left their mark: the great notched stumps which speak of axemen roaming the forest in search of tall straight timber for Melbourne's buildings.

By the light of day a new picture emerges. A sinuous path winds its way in great sweeps amongst the varied vegetation, crossing and recrossing the bubbling water. In the curves of the creek man-made animal homes shelter a great variety of Australia's wildlife. Here at the Healesville Sanctuary creatures great and small, furred and feathered, sleek and scaly, give us a glimpse of the infinite wealth and variety of nature, its spectacle and subtlety.

Yet the first enclosures along the creek were not built for a Sanctuary. They housed the animals needed for observation and breeding for the Institute of Anatomical Research. This was set up in 1920 on 78 acres of Crown land that had once been part of the Coranderrk Aboriginal Reserve.

In the 19th century many Aborigines were displaced by the pastoral claims of the settlers so areas were set aside for them. One of these was a reserve of about 4850 acres near Healesville, stretching from the Yarra River to the slopes of Mount Riddell. The Reserve was called Coranderrk after the Aboriginal name for the Christmas Bush which still grows in profusion along the creek.

As the Aboriginal population of the reserve had declined over the years, part of the Coranderrk land was divided up for the village settlement of Badger Creek during the depression of the 1890s. It was on a small area that had not been cleared for farming that Dr Colin MacKenzie set up his Institute.

He had applied to the government for land not far from Melbourne to set up a field station to study Australian fauna. His particular interest was the contribution that the knowledge of the structure and function of Australian animals could make to medical science. He was given the lease of the land for a nominal one shilling a year providing he paid all the expenses of fencing and running the property.

Two trained taxidermists were employed to prepare specimens and as the fame of the Institute spread world-wide many scientists came to visit the 'Research' as it was known locally. Dr MacKenzie had one of the finest collections of fauna specimens in the world and handsome offers were made for it by American institutions. But he was determined that it should stay in Australia.

In 1924 MacKenzie presented his collection to the nation and the gift was acclaimed as the greatest ever made for public purposes. He made one stipulation: that a worthy place should be set aside to house the exhibits. So the Institute of Anatomy was built in Canberra and Dr MacKenzie was appointed its first director.

When MacKenzie moved from Badger Creek in 1927, he asked the government to declare the land he was vacating, and the adjacent Coranderrk bushland, a National Park. But the idea languished, as did the land, for the next two years. The Lands Department reported in 1929 that the property was deteriorating and recommended that it should be reserved 'for public purposes' under the control of the Healesville Council.

Dr Colin MacKenzie (1877–1938)

William Colin MacKenzie was a man of exceptional ability and originality of thought. The third of four sons of a Kilmore schoolmaster, he won a scholarship to Scotch College and graduated in medicine from the University of Melbourne with First Class honours in 1898.

While resident medical officer at the Children's Hospital around 1900, and in his own orthopaedic practice from 1905, Dr MacKenzie saw many young patients with infantile paralysis and realised the terrible physical disablement and lifelong unhappiness it caused.

MacKenzie became convinced that more could be done to prevent such deformities. He sought to restore muscle function in affected muscles by a combination of rest in the acute stage followed by muscle re-education. He came to the conclusion that the best treatment for neuromuscular diseases could only be discovered by studying the evolution of the muscular system.

This provided the starting point for his long and detailed study of Australian animals. Watching the muscle action of a climbing koala inspired the most famous of his mechanical devices, the arm abduction splint which came into universal use.

'It has restored to civic usefulness thousands who otherwise must have lived and died as cripples...,' wrote Ambrose Pratt in his tribute to Dr MacKenzie. 'He supplied scores of helpless waifs with costly splints and boots and bandages. He nursed and taught these waifs himself.'

His thinking was too revolutionary for many of his professional contemporaries but his unortho- dox treatment was very successful at the Children's Hospital. Patients were soon flocking to his doors. He gave up six hours a day to private patients and six hours to providing free treatment for poor patients.

During the First World War he set up a rehabilitation unit in London, using his skill to restore health to many paralysed soldiers.

Convinced of the importance of the study of Australian fauna to the future of medical practice, on his return to Melbourne in 1919 Dr MacKenzie set up the Australian Institute of Anatomical Research. He aimed to make Melbourne the world's centre of research on Australian animals.

A year later, in his request to the government for an area of land as a field research station, he wrote that: 'Australia has done little to complete the natural history records of her own animals or to work out their structure and function from the point of view of their medical importance to the human race: for without doubt in these animals are to be found the solution of many problems still defying physicians and surgeons...'

On the land he was granted at Badger Creek (where Healesville Sanctuary now stands) he expanded his work. The collection of specimens of Australian fauna he donated to the nation in 1924 numbered nearly 10,000 exhibits. This gift led to the creation of the Australian Institute of Anatomy in Canberra, completed in 1931.

It was the fulfilment of MacKenzie's ambition to establish a museum to house his dissections of Australian fauna, together with a research institute focusing on the development of Australian children. MacKenzie was director of the Institute until ill-health forced his retirement in 1937.

His published works on Australian fauna and comparative anatomy received worldwide recognition. In 1929 Dr Colin MacKenzie was knighted for his services to Australian zoology and medicine.

In ending his tribute to Colin MacKenzie, Ambrose Pratt wrote that: 'He was undoubtedly a good scientist – but he was something rarer still, a great and good man.'

Sir Colin MacKenzie at the gate of the Sanctuary named in his honour

Some of MacKenzie's exhibits at the Institute of Anatomy in Canberra

In March 1930 MacKenzie visited the Reserve with the Shire Council Committee and it was agreed a caretaker was needed to clean up the property and repair the fences. MacKenzie suggested that the leftover pens should be stocked with animals with a view to the formation of an Australian zoo as it would be a great attraction and become a valuable asset to Healesville. He advised the Council to employ William Elverd, a man Sir Colin knew could handle animals as he had collected possums for MacKenzie's research.

The Elverd family moved from their home in Warrandyte to Badger Creek in April 1930. Elverd's son, Bill, writes that the Sanctuary was thick bush except for the home area. There was plenty of work cutting paths and building tables and seats. They caught possums and wallabies. Elverd asked young Aborigines to climb the trees and shake the gliders down, so that he could grab them. His son notes that 'Dad would throw a bag over them – sometimes got bitten.'

There was a lot more water then in Badger Creek and it was easy to see platypus, trout and crayfish. 'The creek flooded every winter and washed the bridge away. After the rotunda was built, we could give visitors strawberries, and cream from our own cow. I remember getting up early to light fires around the strawberries to lift the frost.'

5

In honour of his research work, the Shire Council named the reserve the Sir Colin MacKenzie Badger Creek Sanctuary. Since 1921 the area, including the Coranderrk bushland, had been proclaimed a Sanctuary for native game. Shooting or trapping native birds and animals was prohibited. Keeping night fishermen out of the reserve was one of Elverd's tasks.

The Council was asked to frame regulations for the Reserve and these were eventually gazetted in September 1931. Entry to the Reserve was to be free except on days set aside for fêtes, sports or holiday amusements. It was a reflection of the times that no sports or games were allowed on Sundays.

Two active local organisations, the Healesville Tourist Association and the Badger Creek Progress Association offered to help the Council with the development of the Sanctuary as they felt it would further the interests of the district. Three members of each association were appointed to the Sanctuary committee, under the chairmanship of Councillor A. Harry Blackwood. Robert Eadie, of the Badger Creek Progress Association, worked tirelessly to promote the Sanctuary and in 1933 he was made Honorary Curator.

Eadie was a man with drive, foresight and abundant energy and deeply committed to the idea of reservations for wildlife. While working as an engineer in South Africa he had been involved with the creation of the Kruger National Park which had attracted visitors from all over the world. Strong in his belief in the importance of the project, he was to play a pivotal role in establishing the Sanctuary.

The gazetting of the Badger Creek Reserve stirred the interest of many city people. There was a growing tide of understanding that many native animals were threatened with extinction and needed protection. The fate of the koala, in particular, was of great concern to many people.

David Stead, co-founder of the Wildlife Preservation Society, writes of the wholesale slaughter of koalas in Queensland in 1927: 'Bad as this occurrence has been, it shall be a great turning point in the people's interest in our wild things generally.' And so it proved.

Eadie writes of a public meeting in the Melbourne Town Hall in 1932 'to induce the government to set aside areas where the fauna of the country could live and multiply without molestation. The Mayor of Melbourne presided and there was an unexpectedly large attendance including a strong contingent from Healesville.' A committee was appointed to go into the claims of the rival districts, Healesville, Kinglake, Monbulk, Mornington and Ferntree Gully, and make recommendations to government.

Badger Creek

Log bridges in sweeping curves give varied views of the creek

Dingoes pause for a drink in their walk around the Sanctuary

Rainbow lorikeets

A stroll amongst the kangaroos

Top: Kangaroo paddock Bottom: King parrot

Eadie was on the committee and, when Healesville's claims were considered, put up a strenuous fight for its selection. 'Local loyalty was not the only spur. Remnants of most of the fauna were to be found here. They were, of course, in greatly reduced numbers but the presence of even a few indicated that before the enemy (man) put in an appearance they must have been very numerous. There is an abundance of forest, water, grass, land and everything else suitable for such a purpose. Also a steady stream of tourists who would contribute to the finance required for the maintenance of such a venture.'

A party from the Zoological and Acclimatisation Society was invited to visit the Sanctuary in the hope that they might help to influence the government. The President, Mr Ambrose Pratt, promised his support as 'The Reserve would be suitable for animals needing a change of air, they being much like humans in this respect, needing periodical changes of environment.' He advised that the native jungle should be preserved, that 'the greatest restraint must be exercised in destroying native trees and undergrowth. Animals would not thrive except in their own natural surroundings.'

In August 1932 recommendations were made to the Victorian government that 2000 acres at Monbulk and 2500 acres at Healesville should be set aside as fauna reservations under a board of management.

The sites were chosen because of their closeness to Melbourne, suitability for breeding fauna and general attractiveness. It was realised too that the effort of local residents would be essential for ensuring success.

The committee pointed out that the Healesville site was a very diversified tract of country, capable of supporting various fauna; that the fauna naturally existing in the district was of considerable variety even at the present time and there was no doubt that almost all the most attractive Australian animals and birds could be accommodated under natural conditions. 'The platypus is at present living in Badger Creek, introduced koalas are doing well, kangaroos and wallabies in considerable variety could be maintained and the birdlife is extremely varied. There is no more suitable site within easy reach of Melbourne for the establishment of a wild zoo.'

At a public meeting in Healesville a motion of support was passed unanimously. It declared 'That this public meeting of Healesville citizens fully recognising that the establishment of a natural reserve at Badger Creek for native flora and fauna will be of paramount value to the community of Healesville, and especially to the State of Victoria, do pledge ourselves to support the committee of management in their plans to make Healesville

Robert Eadie (1863–1949)

In the years of the depression in Victoria in the 1890s, many young men set out to seek their fortune further afield. Among these was Robert Eadie, who left his home in Sunbury in 1896 and set sail for South Africa. For the next 26 years his life was to be in the Transvaal, working as a mining engineer.

During the adventurous years of the Boer war, he was one of the party that hid Winston Churchill down a mine and helped him escape to British lines. After the Boer war, he took on the challenge of restoring a badly damaged mine in Witbank and set up his own mine, the Station Colliery.

Later he was awarded the MBE for his voluntary work in many aspects of community life. During the first world war, he was chairman of the Governor-General's war fund. In 1914 he was elected the first mayor of Witbank, an office he held for seven years.

Eadies' wife, Eliza Jane, was also a tireless worker for the community. When she became ill in 1922 he brought her back to Australia and bought land at Badger Creek. Here they involved themselves in district life, and were largely responsible for the erection of a public hall. When he became Curator of the Sanctuary, she supported his work, entertaining an endless stream of visitors, anyone who might help the Sanctuary.

It was in South Africa that Eadie had become deeply interested in the idea of reservations for wildlife, having been associated with the development of Kruger National Park. This background led to his active involvement in setting up reservations for wildlife in Australia. He believed this was essential if the animals were to be preserved for future generations.

His book about his platypus, Splash, called *The Life and Habits of the Platypus*, was circulated world-wide and his observations added a lot to what was known of the platypus at that time.

He commented that: 'Pioneering work has always appealed to me. When finished I am pleased to hand over to other hands and look around for some other work which will demand initiative and perhaps hard work.' Marion Key, Robert Eadie's granddaughter, writing in South Africa in the 1960s, describes him as 'a man of integrity and courage whose example provides a challenge to the citizens of the Witbank of today'. It provides no less a challenge for Victorians in the 21st century — for the conservation cause for which he fought so hard has still many battles to be won.

Robert Eadie with a yellow-tailed black cockatoo

Australia's foremost national park by rendering the committee every assistance, both financially and otherwise.'

Little did they realise what a struggle it was to be. A collection box was in place for donations from visitors to the Sanctuary and at the meeting the Sanctuary receipts were listed at £61/6/- and expenditure £94/13/8, leaving a deficit of £33/7/8.

Over £18 had been used for a new aviary but new shelters for the animals would always be needed and the gate money would not even cover the animals' food. Such a pattern – the need for funds far outstripping funds available – was to put a rein on the Sanctuary for many years to come.

But optimism and enthusiasm were high. Mr Clapp, of the National Travel Association, was convinced that hordes of tourists from overseas, many arriving by ship at the port of Melbourne, would be clamouring to see Australian native life: 'There will be such a number of visitors to Healesville that cars will be end on end and there will be unbounded free publicity. It was for the people to stand by the Committee and it was ultimately intended to get the 2500 acre reserve...'

That hope was never to eventuate.

In 1933 a government subcommittee visited all the suggested sites and in August that year the State Government 'approved of the establishment of two reserves near Melbourne as special sanctuaries for native fauna… One of the reserves will be of about 500 acres at Badger Creek.' While the Sanctuary was given £400 from relief funds provided by the Employment Council, Ferntree Gully, chosen instead of Monbulk, received £600. It was little enough but 'further assistance in establishing the reserves will probably be given later.'

In reality Badger Creek Reserve was given neither the land nor the money that it needed. No move was made to increase the area of land from its original 78 acres. The grant merely paid for a gang of unemployed men to do some clearing and burning off before the summer season.

Saving Victoria's native animals was left up to the people. It was they who embarked on 'a great and interesting work, popular and scientific', the Healesville Sanctuary.

2

Building the Ark

'Out of the darkness of the balcony rose a tall figure who stated that he had 13 sheets of corrugated iron that he could give… Others followed the good example and we slowly but surely forged ahead.' From such meetings in the Healesville Memorial Hall, the wire and the tin, the nails and the hammers were gradually gathered together and a start was made.

As the income was minimal, much of the work had to be done by the committee. This included Shire councillors, but many of them were unenthusiastic, believing that the Sanctuary would be a liability rather than an asset to the district. Writing in 1944, Robert Eadie looks back at the early years: 'I am satisfied that had it not been for the fine work of the advisory members, there would have been no Sanctuary today… So keen were they that on numerous occasions they gave up their weekly half-holiday and walked to the Sanctuary to give it a clean-up. Their work is forgotten by most people but there are a few who will ever remember the unselfish and highly useful work done by them when things were in the making.'

Eadie's engineering skills proved useful. With the help of volunteers he built both animal enclosures and a water-race leading from Badger Creek to create a series of ponds for ducks and swans and later pelicans. These water-bird ponds were a distinctive feature of the Sanctuary for 60 years until they were drained to make way for the new platypusary.

At first there was little more than 'Some of Fitzroy Gardens' surplus possums, various lizards, a goat cart with goats and a few cockatoos that had worn out their welcome and the household furniture! …The first real contribution were two koalas from Phillip Island. They were forwarded by Mr F. Lewis, Chief Inspector of Fisheries and Game, a good friend of the Sanctuary.

We had to report on the condition of those koalas every 21 days – until the day came when I reported on 3 instead of 2. No further reports were required. Just after this came a donation from the Royal Zoological Society, through Mr Ambrose Pratt. Four large cases were received, two fully grown emus and two kangaroos.'

The young koala was a great attraction and after publicity in the Melbourne papers hundreds of people came to visit. 'Mr Elverd was so rushed with enquiries as to where the little bear [sic] could be found that the Shire President had to relieve the congestion by personally conducting many visitors to the trees in which the bears reside.'

Few people at that time had seen a platypus in the wild and it was decided to try to exhibit one to visitors. The Tourist Association contributed £20 for the cost of the materials and a platypusary was built in a curve of the creek.

The appeal of a platypus lay in the air of mystery that surrounded an animal that was duck-billed but furred like a mammal, laid eggs but suckled young. Harold Clapp, head of the National Travel Association, commented that if it were possible for the public to view the timid nocturnal platypus, he could fill a liner with Americans anxious to see this world curiosity in an Australian setting. Eadie was aware that no-one had managed to keep one in captivity for long, but quite how difficult it was going to be was not yet realised.

A platypus was installed in the platypusary in October 1932. Called Glennie (after Eadie's home, Glen Eadie) he seemed to settle down well. In a few days he made his own burrows in his island home. He was offered a variety of food including earthworms, finely cut rabbit, wood grubs, shell-back snails, frogs and tadpoles. But only the worms and tadpoles were eaten in any quantity. Elverd found the animal 'a nightmare' to look after, requiring at least two hours a day.

Everyone was surprised by the animal's appetite: a quarter of his own weight in worms and tadpoles every day. Even though the local residents rallied to the cause, finding 500 worms a day proved increasingly difficult as summer progressed. Very little was known about the needs of a platypus in captivity, so although Glennie only lived for 6 months in the platypusary, at that time it was an achievement to keep a platypus at all.

Many people visited the Sanctuary in the hope of seeing the platypus, but the time of his appearance was impossible to predict. Elverd trained Glennie to appear and take food from his hand when a dinner bell was rung.

But the platypus was naturally shy of crowds. Though sometimes he would appear around 1pm, it might be 4 or 5pm before he emerged from his burrow, so visitors often went away disappointed.

Eadie decided to design a different enclosure that would enable a platypus to be shown more easily to visitors. 'I had visions of taming a platypus and showing it at close quarters.' He built the platypusary, with a water tank, tunnels and sleeping quarters, at his home near the Sanctuary.

A young platypus, found wandering in a patch of maize, was brought to Eadie in February 1933. Eadie writes that 'For ten days and ten nights he refused any food. I was on the point of releasing him when on the eleventh night he had consumed all the food I had placed for him and never looked back…

He was christened Splash and few animals have gained such a worldwide reputation. He was not only a platypus in captivity but a public performer. He would play with me and delight relays of visitors for two or three hours at a time. He would come out of his sleeping quarters when he heard the well-known whistle, and roll on his back on his platform while I nuzzled him with my hand. He was fond of being heaved from end to end of his tank while he clung to a long-handled mop.'

An American journalist wrote that 'the clever and loveable creature went through his act with all the egotistical zest of the most popular vaudeville performer on the bill.' Splash became a world-renowned personality with his own fan mail. People came a long distance to see him and many tourists visiting Melbourne by ship took the trip up to Badger Creek.

Splash astonished everyone by the amount he ate. His daily diet consisted of about ten ounces of worms, a hundred tadpoles, a handful of grubs and two eggs, preferably duck. Many experiments were made to tempt his appetite. Mrs Eadie developed a kind of egg custard that Splash consented to eat after his ration of worms. Over his lifetime he consumed nearly three-quarters of a ton of worms, grubs and tadpoles and over two thousand eggs.

Eadie comments that the success of Splash was a factor in making for the success of the Sanctuary. 'In the early days, numbers of people came straight to my home and took photos and notes. Later I made it a rule to meet people at the Sanctuary and so obtained admission money.' More than 13,000 people from many countries visited Splash during his four years at Badger Creek.

Eadie's observations on the platypus were of interest to both scientists and the public. It was he who discovered that a platypus sometimes has periods of hibernation during the winter months. He showed that with patience and perseverance a platypus may be tamed until it is as docile as a kitten. When Splash died the Sydney Sun commented that Eadie had proved the platypus to be almost as intelligent as a dog and twice as hungry as a wolf.

Splash came to the Eadies as a baby in a tin and four years later left them in a tin – his body consigned to the Institute of Anatomy in Canberra. But this was not to be his final resting place. Winston Churchill was fascinated by the platypus and years later Splash's mounted skin was presented to him by Dr H. V. Evatt on behalf of the people of Australia.

The Sanctuary platypusary was occupied by various platypus after Glennie but one escaped and three others disappeared so permission was refused at that stage to make further attempts at exhibition.

Lack of money was a constant problem at the Sanctuary. The caretaker's wages were paid with money from 'judicious clearing and the sale of firewood'. As the number of birds and animals on display grew, so did the food bill. Money had to be found somewhere. So although the regulations had made entry to the reserve free, in 1934 a request was made to the Lands Department that a small charge should be made. Though many local people supported the reserve, there were inevitably many who were sceptical.

'Sixpence for adults and threepence for children. The first day was looked to with hopeful interest on one side and scarcely disguised cynicism from those who had little or no faith in the venture. The day for collecting the silver coin arrived – and so did the rain. Six adults and three children turned up – only 3/9d but a beginning. We purchased 12lbs of wire nails with the amount and those nails are holding together a good many of the erections in the Sanctuary today,' writes Eadie ten years later.

The official opening of the Sir Colin MacKenzie Sanctuary for Australian Fauna and Flora took place on May 30, 1934. It was planned to capture publicity and hopefully the sympathy of those who could help the Sanctuary. The Hon. W. H. Everard MLA presided over a lunch in Badger Creek Hall for a hundred guests including several cabinet ministers. Everard was the local Member and a strong supporter of the Sanctuary. Over many years through the 1930s and 1940s he lobbied his parliamentary colleagues on its behalf.

In declaring the Sanctuary open, the Chief Secretary, Mr Macfarlan, said that the area was far too small and that it would be desirable for Coranderrk land to be transferred to the Sanctuary. But the Aboriginal Board needed the money received for grazing rights to support Lake Tyers Aboriginal Reserve.

If the Minister of Lands could provide the Aboriginal Board with alternative land…? Mr Dunstan, the Minister of Lands, agreed more land was needed and felt sure the matter could be speedily resolved.

Yet for all the talk, despite requests and deputations, letters and publicity, nothing was to happen. Eadie notes that 'In the early days the local committee made great efforts to obtain security of tenure. We pointed out that we were doing work for the State without reward but we might as well have talked to the magpies for all the effect it had…Many of our MPs regard our irreplaceable animals as so much meat and leather.'

The Sanctuary Committee had many other problems to contend with. The stream that gives the Sanctuary its character and charm can change from a trickle to a torrent at any time. The year the Sanctuary was opened the creek broke its banks and coursed through the Sanctuary with unbounded ferocity. In December 1934 the local paper reported that: 'Badger Creek turned into a raging torrent that swept through the Sir Colin MacKenzie Sanctuary, carrying nearly everything before it and doing incalculable damage… The Sanctuary improvements were nearly all washed away. The recently-established museum, wherein were housed many valuable exhibits of mounted specimens of native fauna, was inundated and most of the exhibits destroyed. The building was saved by a gang of workmen who, neck-deep in the raging torrent, managed to anchor it to a giant gum tree. The Sanctuary kiosk was also flooded and a quantity of stock and souvenirs, together with many furnishings, destroyed. This building was also saved by being anchored to a tree. The extensive fence surrounding Kookaburra Park was swept away and the kangaroos, emus and angora rabbits are now at large in the bush'.

But the creek in its quieter moments served a useful role as a cooler for the kiosk's soft drinks. Running the kiosk can have been no easy task for as well as having initially neither refrigeration nor hot water, the lollies and perishables had to be taken home at night because the possums would squeeze between the trellis and help themselves. The kiosk was run by Miss Moreland and her sisters. They cooked at home and carried cakes and scones to the kiosk in a suitcase. Water was boiled in a big urn on the open fire.

Though some animals were in captivity so that they could be shown to visitors, many of the animals were free to wander about the park. Crosbie Morrison, well-known naturalist, broadcaster and editor of the magazine Wildlife, writing in 1936, saw the future of the Healesville Sanctuary as similar to that of Whipsnade in England. There 'Animals roam under conditions which resemble as closely as possible their native haunts. The Victorian sanctuary in the hills would be an improvement on this, for the surroundings of the animals would not be counterfeit but genuine, and as none of the native animals of Victoria is essentially vicious and dangerous, restraint could be reduced to a minimum…At Healesville there are a few enclosures for specimens of smaller animals which might never be seen if they wandered at will in a large area of open forest, but many of these animals are at large. There is also a flight aviary for specimen native birds, but most of these are free to come and go in the open forest, retained in the Sanctuary only by the attraction of abundant food.'

The zoo in the hills was also seen as a potential holiday home for animals from the Melbourne Zoo, as koalas were proving difficult to keep healthy in captivity. Morrison adds that 'At Healesville the koalas are really not in captivity at all, and they remain healthy', so the zoo animals, 'Having taken their turn on exhibition, are returned to Healesville and their place is taken by another squad from Badger Creek Sanctuary.'

But the koalas proved more ravenous guests than had been anticipated. Two large manna gums were expected to be enough for each animal but it was soon found that the food trees quickly became very bare and the koalas had to be moved to another area of the park to give those trees a rest.

In the 1930s many articles appeared in the newspapers with pleas for koala sanctuaries. Due to hunting, the ravages of bush fires, the spread of European settlement and the fox, Victoria's koala population had been reduced 'in less than half a century from several millions to a few hundred…the koala must utterly disappear from Victoria before another decade has elapsed unless the right measures are taken.'

At Healesville 15 female and 5 male koalas from French Island were flourishing but the need for more land for the Sanctuary had become urgent. Mr Lewis, the Chief Inspector for Fisheries and Game, recommended to the Chief Secretary, Mr Bailey, that at least 300 more acres of Coranderrk should be added to the Sanctuary. However, the Aboriginal Board remained reluctant to give up Coranderrk land. The Chief Secretary did not press the point because he took the view that the provision of a completely natural environment for the koala was more important than establishing a sanctuary in a location easily accessible to the public. Others believed that the best future for the koala was in places where they could be defended from fire and the fox.

The Sanctuary has always had to walk a fine line between being a sanctuary in the sense of a place devoted to the conservation of Australian animals, and being a showplace where people could come and enjoy these animals and learn more about them. The fact that so many Australian animals are nocturnal makes it a reality that when left entirely in their natural state few people will ever get the chance to see them.

Though for the founders of the Sanctuary the conservation of the fauna was the guiding light, they accepted the fact that visitors would provide the funds needed to carry on the work. They also believed that the more opportunity the public had to appreciate native animals, the more they would strive to protect them.

But what would attract visitors? On that point the views diverged. In 1937 the issue of snakes in the Sanctuary divided the Healesville community, for while some felt they would be a great attraction, others feared that they would scare people away from the district. To Eadie 'Snakes are beautiful creatures and will make first-rate exhibits in the reserve.' To Shire President Mowle, and he was not alone in his view, 'They are repulsive, dangerous vermin, and should be exterminated.'

This row brought to a head the disagreements between Eadie and the Shire President and Eadie resigned from the curatorship of the Sanctuary. The Council took the opportunity to dispense with the services of the advisory members altogether.

Some in the Council wanted to appoint David Fleay, already well-known as a naturalist from his work in charge of the Australian Section of the Melbourne Zoo, as the new Curator. But Mowle and others felt the expense was too great, that it would be an unwarranted drain on the Shire finances. This was understandable because many in the Shire felt that their rates would be better spent on roads and gutters.

Although visitors to the Sanctuary increased every year, reaching 16,000 in 1936, the expenditure of £751 already outstripped gate receipts of £626. Where was the salary of £350 to come from? The Chairman of the Sanctuary Committee, Cr. W. J. Dawborn, believed that Fleay himself would be the means of attracting funding and that 'If we falter, some other progressive tourist resort will brush us aside. We have an opportunity now and on no account should we let it slip.'

As the animals were actually under the guardianship of the Fisheries and Game Department it was thought possible that Healesville might lose the Sanctuary if a qualified curator was not appointed. So in October 1937, despite the doubters, David Fleay was invited to take charge.

Soon after, Fleay moved in – and so did over 100 snakes! The skills he brought to the Sanctuary were just what was needed for this fledgling sanctuary to take flight. To sustain the momentum, a man of passion and practicality, a thinker and a doer, was needed: and in David Fleay the Shire Council was lucky to find such a man.

But where to house these wriggling, writhing reptiles? 'There is no snake pit in the Sanctuary, and on the reptiles' arrival, in boxes and bags, the Sanctuary attendants were sorely tried in liberating them and in finding safe quarters for their lodgement. Eventually they were placed in the empty platy-pusary, but several of the more lively specimens evinced an easy ability to scale the walls, with the result that wire netting barricades were hastily improvised.'

However Fleay's cargo of snakes were a stunning success. Instead of driving people away the snakes proved an irresistible attraction, 'resulting in an attendance of thousands instead of hundreds weekly.'

When the new snake pit was opened in July 1938, Mr Gale, chairman of the State Tourist Committee, commented that 'The Sanctuary has been a means of cultivating a national pride and sentiment in Australian species of the animal kingdom. Under David Fleay the originality and attractiveness of this natural zoo is continually being enhanced and already it is the largest and most popular institution of its kind in the Commonwealth.'

The Sanctuary committee at the opening of the new snake park

David Fleay (1907–1993)

Few men have contributed so much to wildlife in Australia. David Fleay's talents lay not only in his deep affinity with the animal world, which led to his discovery of so many of its secrets, but also in his enthusiasm for sharing his knowledge. Through his writings, or his chats about animals while holding a platypus or flying an eagle, he imbued thousands of people with a love for the natural world.

His sister Mary Beasy recalls how David's passion for wildlife started at an early age. 'His pets invaded our house and our lives. He borrowed Mum's prize flour bin to catch a wombat and I remember how wild she was when he brought it back covered with air holes so the wombat could breathe.' She remembers David returning from a visit to relatives and 'Mother opened his case to find all his clothes gone and a large goanna and tortoise eggs in their place. A month later we found his clothes in a kerosene tin sitting in the station goods shed!'

It was his mother, the talented artist Maude Glover Fleay, who encouraged his interest in the animal world. Finding he was not cut out for a career at his family's Ballarat pharmacy, David went on to study zoology. His great interest was in the life studies of the animals of the bush, but most of the scientific information available related to stuffed specimens and the classification and distribution of animals.

It now seems amazing, but Fleay's request to his Professor of Zoology at Melbourne University in 1931 that he do post-graduate study on living animals drew the reply 'There's no room, no value and no sense in work of that nature.'! So David Fleay left university, married Sigrid Collie, a fellow science graduate, and joined the Education Department. Though an excellent teacher, he yearned to learn more about native animals. In 1934 he resigned to design, build and stock an Australian fauna section at Melbourne Zoo.

Fleay's wholehearted devotion to the welfare of his animals caused difficulties at the Zoo when he refused to feed the insectivorous frogmouth on horsemeat. His dismissal caused a furore and he was offered reinstatement, but by then the allure of Healesville's forest-clad hills was to prove too great. Fleay moved to Badger Creek in 1937 to take up the Directorship of the Healesville Sanctuary. The Zoo's loss was undoubtedly the Sanctuary's gain and the move gave Fleay the freedom he needed to explore the ways of the wild.

Recording life histories, in pictures and writings, of the lesser known species of Australian animals was his primary aim. During his years at the Sanctuary he achieved the first captive breeding of the platypus and many other species.

Fleay had started extracting venom for the Commonwealth Serum Laboratories' production of anti-venene in 1927. His demonstrations of milking tiger snakes and copperheads at the Sanctuary were very popular. Fleay was bitten by a tiger snake in 1940 while he was examining it for parasites. He was driven at break-neck speed to hospital only to find that, because of the war, no anti-venene was available. Fleay milked the first taipan in 1950, after it had killed a young Queenslander.

In 1952 he moved up to Queensland to set up his own Fauna Centre where he continued to display and breed animals and to increase our understanding of their behaviour and their needs. His depth of commitment to his task is highlighted in his success in finally breeding the powerful owl after trying for thirty years. David Fleay refused all offers from developers for his valuable Gold Coast property and in 1984 he passed the Centre over to the Queensland National Parks & Wildlife Service to ensure that it was preserved for all time.

Among Fleay's works are countless articles, scientific papers and several books on native fauna. He was also co-founder of the Wildlife Preservation Society of Queensland. His contribution to science and to conservation was recognised in many different awards over his lifetime and included the Natural History Medallion in 1941, the Order of Australia in 1980 and an Honorary Doctorate of Science from Queensland University in 1984.

David Fleay with six-week-old powerful owlets, collected from Mt. Riddell. The young male is on Fleay's shoulder and the female on his hand

Dr Colin MacKenzie's original house from the 1920s. Sanctuary staff lived there for the next 60 years

Colin MacKenzie

Bill Flverd showing a ringtail possum – early 1930s

Robert Eadie feeding a brushtail possum

Some members of the government committee which inspected the Badger Creek site in May 1933. Boomerang throwing was part of the Sanctuary scene in the 1930s and 1940s

Official opening of the Sanctuary on 30 May 1934

The waterbird ponds created in the 1930s by diverting water from Badger Creek

1934 floods race through the Sanctuary. The kiosk and museum were nearly washed away. The museum, on the right, housed a variety of exhibits from stuffed animals to specimens in jars

Jimmy with Digger, a popular Sanctuary character

Collecting snakes along the Murray River

A new batch of snakes in the Sanctuary snake pit

Vern Mullett collecting birds for Sanctuary stock from Kow Swamp, northern Victoria

An enclosure for yellow-bellied gliders, built in Fleay's 'spare time'. Young trees were often used as uprights for the first enclosures

One of the early enclosures. It housed Tasmanian devils and, later, wombats

Garrie the goanna lived in a hollow 60 ft up in the lizard enclosure. A keeper had to be hauled up with a rope to check on his welfare

Adult water-rat

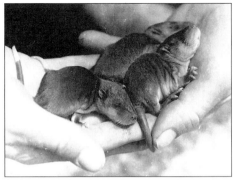

The first water-rat young to be born in captivity

Fleay's Hydromussary, or home for the water-rat

J.R. Goodisson with three barking owlets in the grey box tree where their nest was found

Among the many successful breedings during Fleay's directorship was a Tasmanian masked owl. This youngster, photographed at about 6 weeks, was christened by Fleay 'The Giant Powder-puff'

Native cat (eastern quoll) with young

This wild kookaburra would often land on Jean Osborne's bike or head as she went around with the meat or fish

This large eel weighing 11 pounds lived with the snakes at the Sanctuary. The eels at Badger Creek began life in the Coral Sea. After drifting down the east coast of Australia they entered the Yarra River. Travelling by night, they swam upstream climbing waterfalls and rapids to reach their mountain home. They live in the creek for 10–20 years and then swim back to the Coral Sea to spawn

Tree kangaroo

A freshwater crocodile

Fleay and assistant Alf Wright outside the Tropical House with a Queensland python

The Queensland striped possum that was flown down from Cairns packed in coconut fibre. It uses its very long fourth finger to hook insects out of crevices. At the Sanctuary it was supplied with witchetty grubs and wood moths by an Aboriginal family from Coranderrk

Pong, a New Guinea cuscus

Frill-necked lizard

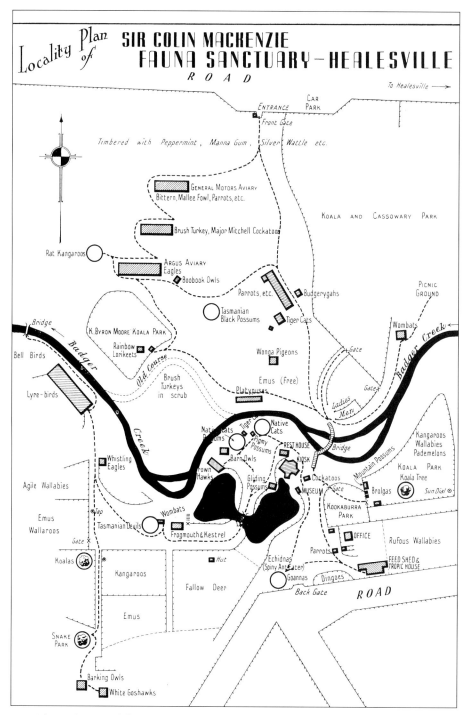

The 1941 map of the Sanctuary shows the variety of animals on display

LEARN TO LOVE OUR OWN NATIVE ANIMALS

To the Tall Gums on the babbling
Badger Creek at the

SIR COLIN MACKENZIE SANCTUARY

HEALESVILLE

is only an hour's picturesque run
from Melbourne

Come and see a large assembly of members of our
furred and feathered fauna. Here you may come
into actual contact with

"Wenda" the Wombat, "Jill" the Platypus,
"Horatius" the Hunting Eagle
"Wylah" the Black Cockatoo, "Billy Bluegum,"
"Garrie" the Giant Goanna
"Ethelbert" the Emu and His Family,
and many others.

At the foot of the mountains eighty acres of magnificent bush,
fern-gullies, bellbird colonies and water-bird lakes.

Don't miss the New and Very Spectacular
SNAKE PARK

"Horatius Attains Film Fame"

During the Summer

DON'T MISS **A DAY'S PICNIC**
at the

Sir Colin Mackenzie Sanctuary

HEALESVILLE - - - VICTORIA

Phone, Healesville 7

*I*T is beautifully cool and shady along
the pretty Badger Creek where you
may meet such outstanding Australians
as Koalas, Lyre birds and the Platypus.
Numerous other furred and feathered
natives include Kangaroos, Wallaroos
and Wallabies, Dingoes and Tasmanian
Devils, Tiger Cats, Emus, Cassowaries,
Brolgas, Tame Wi'd Parrots and a
great variety of other birds.
You will be delighted to wander
through this bushland reserve meeting
scores of friendly animals and birds.

See the tropical House with huge Brown and
Small Green Pythons, Tree Kangaroos and
Cuscuses.
Large Open Air parks house the venomous snakes.
Spotlight tours by night by special arrangement.
Refreshments and Hot Water available at the
Kiosk inside the Sanctuary.

Attracting people to the Sanctuary in 1940

The numbat, brought over from Western Australia in 1941, lapped up 10,000 termites a day

US Red Cross personnel make friends with a koala. During the war, and in the early post war years, many Americans visited the Sanctuary and often contributed generously to its upkeep

3

In the early days the animals came to the Sanctuary in many different ways, some from the wild and some from other organisations. Some were given and some were swapped: like the emu and pair of swans that came from the Melbourne Zoo in return for some Sanctuary logs.

The early directors went on regular collecting forays to many parts of Victoria and further afield. Eadie writes of one of the most adventurous expeditions which coincided with the disastrous floods of 1934. He had obtained permission to remove several koalas from South Gippsland, but by the time he and his assistant, Jimmy Harris, arrived, the rain was falling in torrents and a gale was blowing. They located two koalas in a tall tree and Jimmy was keen to try to get them, even though the slippery surface of the bole made it dangerous. As he approached the first koala it moved out onto a branch which was swaying alarmingly.

Eadie called to Jimmy to come down but he pretended not to hear. He had a rope over his shoulder with a bag on the end. Eventually he got within reach of the koala and slipped the noose over its body. After a great fight he got the animal into the bag and lowered it to the ground. The second koala was captured the same way. 'They are strong creatures and when their claws take a grip it is not easily broken. To have to battle against wind and rain and at seventy or eighty feet from the ground with the branches swaying ten to fifteen feet is a performance few could have accomplished.'

They found other koalas not so high up and left South Gippsland with nine. The whole countryside was flooded and twice they had to be pulled out of the water. When they reached Melbourne three days later they found that they could not get to Healesville so they liberated the koalas in a roomy shed.

A night trip out to Heidelberg was made in search of food for the koalas and the gum tips they brought back were readily eaten. One young koala died but, when the floods subsided, eight were finally brought to their new home. This was the beginning of a large koala population at the Sanctuary. By the late 1930s there were 50 koalas in the colony and 15 females had young in the pouch.

The platypus was equally difficult to catch. Reg Stanley, President of the Healesville Tourist Association, spent many nights lying on a wet, slippery log in the river, waiting for a platypus to come close enough to pick up by the loose skin which surrounds the body. As the platypus is very wary he could not move even when badly bitten by mosquitoes!

Many members of the public simply consigned animals to Eadie or Shire President Mowle in the 1930s. Mowle caught a number of animals himself including a giant wombat.

The goat was given by Dame Nellie Melba's brother, Charles Mitchell, 'for the purpose of clearing blackberries and other noxious weeds from the reserve in preparation for the centenary. Nanny, as this high-pedigree Angora is known, is proving equal to the job and has settled down with relish to the task of blackberry eradication.' He later donated 'a very natty little goat cart as well. Excursions at the Sanctuary by this means of transport will soon become highly popular...'

Some animals were given by other zoos. David Fleay:

In 1936 we had sent Mr Eadie two young Brush Turkeys from the Melbourne Zoo, in the belief that they were a true pair. However, unfortunately it became evident as they reached maturity that both birds were males. In the following year, being exported myself from the Melbourne institution to Healesville, it seemed to me that the beautiful bush and scrub of the Sanctuary were ideally suited to the well-being of free-living Brush Turkey and with that end in view we tried far and wide to obtain a hen bird and also a suitable large rat and fox-proof aviary in which to breed young birds that could be released.

After eight months the Adelaide Zoo gave us a very old and cross-billed hen, the only one they had. In the meantime two generous brother-members of the Victorian Field Naturalists' Club had contributed £100 which built the fine chain-wire enclosure named after them – the W. H. and M. J. Ingram Aviary. Badger Creek school donated all its autumn oak leaves and so the finer of the two cock turkeys built his first cock mound in August 1938.

Half a dozen chicks emerged from that mound during the late spring and summer – the old deformed hen had done her work well. The youngsters grew into healthy tame birds and knowing of the menace of foxes, goshawks and domestic cats, it was with some trepidation that we released them at the half-grown stage. Three disappeared but the others stayed and the first really big and self-chosen incubator-mound of a tame-wild bird made its initial appearance in August 1939, in thick creek-side scrub down near the Byron Moore Koala Park.

From the six or seven mounds now (1946) in the Sanctuary, chicks emerge each spring and summer, and though there is a certain mortality, the species has taken a firm hold in the locality.

Fleay made regular expeditions in search of animals and a favourite place was Moira Lakes near Barmah. Writing of an expedition there in 1938: 'The snakes are there in hundreds, crawling in and out of rabbit burrows, basking on logs in the sun or taking the waters at the lakes edge which provides food and drink both, for the water is swarming with frogs...A fine fat black snake was our first encounter...Cornered, he fought savagely, rearing half his length off the ground, flattening the short hood along his neck, and flinging himself at us with explosive hisses, while the forked tongue flickered angrily.' The snakes were grabbed by the tail and dropped headfirst into a bag. The party captured ninety snakes on this trip – mostly tigers which were plentiful along the water's edge.

Robert Fleay remembers one expedition when he accompanied his father. The bag of venomous snakes had been stowed in the car for the return journey but when David Fleay opened the car door to climb in, he found a very large snake on his driving seat! The snakes had escaped from the bag so the car had to be almost taken apart to find all the snakes before it was safe to set off home!

Birds were added to the Sanctuary collection as opportunity offered. To capture a brood of young rosellas, wire was put over the nest hole. The parents fed their offspring through the wire and once the young were fully fledged, they were brought to the Sanctuary.

In 1935 Fleay captured a young sea eagle by firing bullets at the limb of a giant red-gum on which a nest had been built 80 feet above the Murray. After 461 bullets, the 14-inch diameter branch crashed into the water, bringing the nest and eaglet down with it.

Horatius, the wedge-tailed eagle that later became Horatia after she unexpectedly laid two eggs, was collected by Fleay in 1937 'after a particularly nerve-wracking climb, from a great eyrie forty feet up in a crazily-rooted eucalypt leaning out from the wall of Coimadai Gorge near Bacchus Marsh.'

Horatia carried on a gauntleted arm in the late 1930s. She proved adept and spectacular at hunting – once actually capturing a large wallaby. Fleay writes that 'Horatia often provided thrills for visitors when flown free in surrounding paddocks, putting on a turn of acrobatics in pursuit of stuffed rabbits towed along at speed by fleet-footed boys. Dramatic diversions occurred in winter when a pair of wedgetails "owning" the area screamed down in power dives trying to chase her from their territory...Falconry, with trailed or swinging lures as the focus of attack, proved a strong public attraction. Peregrine, black and little falcons, whistling kites, Australian goshawks and little eagles all took the stage in turn.'

Horatius/Horatia

Horatius dominated the Sanctuary scene for a decade, a star exhibit and photographic model whether guarding the bridge over the creek or planing over the paddock.

Each year Horatius built a nest as high as possible in the aviary. But, at the age of thirteen, 'he' surprised even Fleay: 'In 1950, I humoured the ambitious bird by spending odd minutes in the days of late July running up a ladder fixed at the nest end and helping her with sticks and leaves. She was most delighted and excited, even beaking my ears and hair with her formidable but amazingly gentle beak. That was the year that Horatius became Horatia for the great bird staggered us by laying two big eggs and reversing all preconceived ideas about her sex. She sat so hopefully and for so long that we got Mr Norman Favoloro from Mildura to kidnap two newly hatched eaglets from a Mallee eyrie and on a dark night, when Horatia could not see what was happening, swapped the babies for her eggs. She reared them beautifully...'

Horatia feeding the eaglets with slivers of rabbit flesh

Many expeditions had their dangerous moments. One wild windy night, when Fleay and Vern Mullett (later to be director himself) were out on Mt Toolebewong, 'A spotlight from a car had picked up what appeared to be a true albino greater glider sitting at a great height in a mountain grey gum. Using a rifle, a number of rounds were fired at a selected spot on the bough behind the beautiful, long-haired creature. But a great gust of wind suddenly snapped the weakened and weighty bough and blew it down, complete with animal, right on top of us.'

In 1939 many victims of the bushfires found succour at the Sanctuary. Fleay notes that overnight enormous tracts of fern-starred mountain gullies and picturesque bushland were converted into charred graveyards.

He found the platypus, Jack, in a tiny pool in Badger Creek which was almost completely dried up. 'We made the discovery at 2 am during a patrol in a smoke-filled, hellishly hot night, when it seemed likely that all our work and hopes might be destroyed by fire.'

Fleay combed the mountains for other victims: 'For weeks after the flames died on Mounts Riddell and Toolebewong we found poor tortured greater gliders crawling weakly over the charred ground, their black colour making it hard to see them amid the desolate silent world of dreary stark trunks and ashes. Our Badger Creek Sanctuary and neighbouring Coranderrk Reserve had been spared the fire, so we gathered up all the victims and liberated them among our own untouched forest. Many recovered – some did not. But with those of their kind already there, we have today a goodly population of greater gliders regarding the Sanctuary as their home ground.'

Fleay comments that wombats came off best in the fires because they had good dugouts for shelter but many died of starvation afterwards. Some wombats were captured and held as temporary boarders at the Sanctuary. Wallabies were often saved by going into a creek, only to have their feet hopelessly burned on the carpet of glowing embers.

A 1939 trip to New Guinea netted an interesting bag including cuscuses, tree kangaroos, pythons, sea snakes, and two little orphan New Guinea gliders that were rescued from an attacking horde of green tree ants.

The gliders lived comfortably in a large box intended for breeding mealworms, perched on warm pipes in the Sanctuary's Tropical House. They entered the box through an opening under the warped lid. But warm weather straightened the lid and they were trapped and died. No one realised at first as the food put out for them each evening disappeared, but it turned out the mice had eaten it.

A python that Charlie Mitchell brought back from Queensland came down on the boat in a box marked 'orchids', just in case some passengers were nervous about having a snake as a travelling companion.

Some acquisitions were planned while others were fortuitous. In 1946 David Fleay caught two water-rats while he was trying to catch platypus for the New York Zoo. The male water-rat was golden-brown with an orange belly, while the female was the colour more common in the Healesville area, grey with white underneath.

And some just came by accident, like the tiger cat in the 1950s that wandered into a rabbit trap in Kiewa. It was sent to the Sanctuary by rail but escaped from its box and was on the loose until it was unwise enough to kill a local resident's laying pullets.

Even in the 1960s collecting was a regular part of a director's life. Vern Mullett made an annual pilgrimage to Kow Swamp, north of Echuca, to collect birds and snakes and 'whatever can be caught'. A rookery for cormorants, spoonbills and ibis, there was plenty to catch and each year twenty to thirty specimens would be brought back to the Sanctuary.

Collecting spoonbills involved climbing up to the nests in the swamp trees to find young chicks that were old enough to be taken. 'Sometimes the chicks will instinctively plop 15 ft or more from the nest and dive below the surface. Swans have to be chased by boat until they are tired. Grab or net them for the box. It may take half an hour's risky boating for miles to out-manoeuvre flapping birds. For ibis you wade through waist-deep water and take young ones from the nests in the reeds. Sometimes you will catch a larger bird, like a crested grebe, with a rugby dive in the tangled morass.'

Breeding from animals already in captivity and so gaining a greater understanding of the animal's lifestyle was always an important aspect of the Sanctuary. The early guide books make it clear that 'the aim is the preservation of all native animals quartered in the Sanctuary with special emphasis on breeding rare species and recording hitherto neglected life histories.' The first platypus breeding in 1943 grabbed world attention, but there have been many other notable firsts over the years.

Many were Fleay's successes and deservedly so for he took immense trouble to provide the animals with everything they needed. When he decided to try to breed from the water-rats, he built a special enclosure which he called a 'Hydro-mussary' from Hydromys meaning water-mouse. This home had a swimming tank, landing shelf and a series of grass-lined wooden burrows. Fed with fish and yabbies, the water-rats flourished and a month later the female produced three young, in a grass-lined nest at the end of a burrow. This was the first litter of water-rats to be born in captivity.

Fleay had his own collection of animals before he came to the Sanctuary. Among those he brought with him to Badger Creek were native cats (eastern quolls), an animal once common in Victoria but already in the 1930s an endangered species on the mainland. 'The discovery of eight little shapeless pink blobs in the pouch of one of our two female native cats caused a con-siderable stir and some rejoicing at the Sanctuary. The parent animals have flourished exceedingly, fed on rabbit carcasses and fish. They share the run with some silver-grey possums and it is no unusual thing to find possum and native cats curled up together for sleep in the daytime.'

From time to time some unusual young have been born. In 1943 a Queensland squirrel glider and Victorian sugar glider interbred, producing one young female, which then had two young fathered by her own parent. 'All members of the 3-generation family are doing well and living together in a great untidy leaf-nest in the floor of a hollow log. Cold weather and frosts, on which nights the animals emerge very little, present no problem to them for they cuddle up in a tightly rolled furry mass, enjoying the community warmth.'

Cockerellas, young from a female sulphur-crested cockatoo and a male corella, had blue patches round their eyes like their father and yellow crests like their mother.

Three barking owlets, found in a nest in a hollow of a grey box tree in the Riverina, were brought to the Sanctuary in 1937 and their dog-like bark-ing calls became a familiar feature. Later another female barking owl was brought in after being hit by a car. Though she laid eggs in a hollow log in the aviary no young were hatched. So in 1941 a special enclosure was con-structed in peppermint bush well away from the public pathways, and she was moved there with one of the males from the Riverina.

The male was soon spending a lot of time scraping a depression in the rotten wood of a large hollow log suspended from the ceiling and here three owlets were successfully raised. But the owl parents became very aggressive in defence of their family, so when Fleay entered the enclosure his armour con-sisted of a thick overcoat, a milk bucket helmet and a long-handled broom!

Fleay writes that during his years at the Sanctuary 'We were able to assemble fairly complete life histories from our observations on a number of little-known species, in particular, the tiger quoll, southern quoll, Tasmanian devil, rat kangaroo, various possum gliders, the Tasmanian masked owl, barking owl, and the fascinating powerful owl, which was finally run to earth and watched during its nesting performances.'

As well as the scrub turkeys, emus, koalas and wallabies were bred under controlled conditions and liberated in the habitat area. 'Some were great attractions, such as Father Emu strolling along with a group of piping chicks, a yellow-collared scrub turkey feverishly scraping up a 1 or 2 ton mound among the tree ferns, and Wau, the New Guinea one-wattled cassowary who came racing through the bush to do a war-dance every time I shot at a rabbit.'

Fleay comments that a dozen emus in eighteen months ate out every trace of common bracken. He found them more efficient than goats at clearing blackberries but 'as a by-product of the fruiting season, blackberry "jam" was somewhat of a drawback on each and every pathway!' He added that 'Emus lose few opportunites when it comes to the disposal of papers, ice cream cups, bottle tops, keys, coins and even brooches!'

The silvan setting of the Sanctuary was a natural home to many species of bird including grey fantails and yellow robins, kookaburras, crimson rosellas and flocks of king parrots. 'Hundreds of bellbirds tinkle in a 3-acre colony of peppermint trees.' In November 1937 gangang cockatoos are recorded perching in scores in the wattle trees of the goanna enclosure and yellow-faced honeyeaters and red-browed firetail finches nesting in numbers along the creek. In the Sanctuary and the Coranderrk bushland over 180 species of bird have been recorded.

In recent years animals are taken from the wild only when absolutely necessary. Animals needed to increase or replace stock are generally sought from other institutions. Some of course are bred in captivity. There is emphasis now at Healesville on breeding animals that are rare or endangered in the wild.

The Powerful Owl

Breeding the powerful owl was a project close to Fleay's heart. So there was great delight when a wild bird answered Ferox's call and then started to sit occasionally on the roof of his cage. Believing the bird to be a female, Fleay turned the roof into a giant trapdoor and soon the unsuspecting bird became a reluctant captive.

The two birds sat apart on their perches in Ferox's cage with no enthusiasm either for eating or calling. After a couple of weeks, for no obvious reason, Ferox (pictured) got sick, and despite being given every care, he died a few days later. Fleay was devastated. He now realised that the new bird was probably also a male and another one calling in the area was presumably its mate. The mate was eventually caught by trapping it with netting on one of its favourite perches high up in a silver wattle.

A new enclosure was built for the pair among standing timber, walled thickly by a teatree palisade. 'Dragged in by draught-horse, the nesting hollow was pulled into lofty position with block and tackle. It had a deep rotten wood interior, being altogether the last word in decoration and comfort for the most diffident of owls. There the pair of dignified detainees dwelt in beauty and grace — my great hope and my despair for three whole years. They ate royally and looked well but might just as well have been zombies.'

When there was still no sign of courtship, Fleay set them free. He came to the conclusion that taking young owls from the wild and rearing them in captivity would be the only possible way to breed them. He finally tracked down a powerful owl nest on Mt Riddell and brought two month-old young to the Sanctuary in 1942. The female, Hookie, shared a large bush aviary with a fine young male for ten years but breeding success still eluded Fleay. He was finally to fulfil his dream in Queensland many years later.

4

Anchoring the Ark

A fallen log, a hollow in a tree, a trunk for a post: these were the raw materials that made homes for the animals in the early days. Eadie and Fleay, as designers, builders and handymen, provided shelter for the animals from the wealth of the bush around them.

As there was no overall plan for the Sanctuary, the enclosures were built wherever seemed most suitable at the time, with whatever materials were to hand. Fleay notes that 'In those days it was start from the ground up. In most cases our building material was split from brown stringybarks.' For a home for the Tasmanian masked owls, the trees themselves were used as uprights.

One of Fleay's reasons for going to Healesville was that he knew the area to be the haunt of the powerful owl. Ferox, the orphan powerful owl he had raised with such devotion, would be in his element. He wrote that: 'One of the first structures was a large sheltered aviary of bush timber and netting, built along a picturesque creek track amid a grove of silver wattles, and here at long last Ferox entered the environment beloved of his kind.'

Some animals made use of the natural homes in the forest. A giant goanna named Garrie lived in a hollow 60 feet up in a tree in the lizard enclosure. To check on his welfare the keeper had to be hauled up the trunk with rope and tackle!

The provision of reserves for the koala was still a very important aspect of the Sanctuary in the late 1930s. Fleay reports that 'six separate koala parks are fenced and nearly fifty koalas inhabit these tracts of timber.' Other koalas had escaped from branches hanging over the fence and were in the neighbouring trees of Coranderrk. 'Altogether the population of the whole exceeds 100. Quite a job of boundary walking involved each day, inspecting the long lines of fencing, filling in great gaps caused by burly wombats scratching away beneath or repairing the damage caused by falling limbs. The fences themselves are low but complicated by a right-angle overhang which effectively stops the wandering habits of the nomadic koala.'

In the 1938 guide to the Sanctuary: 'A few koalas are kept in an exhibition yard where they have a wattle tree to climb and gum leaves to eat. They are exchanged frequently and no koala is kept out of the forest for more than a little while.'

The guide makes it clear that for the most part the animals were left to live out their lives as naturally as possible:

Wenda, the Flinders Island wombat, came with us on our walk, over the little bridge between the two lakes of the waterbirds. Here the graceful black swans, the magpie geese, the white ibis, the trim and stately spurwinged plover, the rare nankeen night heron and the smooth and dignified Cape Barren geese live in harmony with the black ducks and teal. All but the rarest of these waterbirds are free to fly but they have made the lakes their home.

From the lakes a paddock extends down into a forest and gully of trees. The black-tailed wallabies, the red and great grey kangaroos live in natural grounds...A lovely native companion (brolga) with a drooping wing is their friend and a pair of dainty fallow deer come down from the forest above and feed at the same trough...

Wallabies, bandicoots and gliding possums are wild in the accepted sense but trusting of visitors...The lizards are 'at home' in a circular yard. The lace-lizard, the blue-tongue, the stump-tail, the bearded lizard and the water lizard all have their natural food. Large white hen eggs lie on the ground for the five-foot-six goanna...

Though whether that would be enough to satisfy Ockie, the six-foot-three specimen that lived there in the 1940s is doubtful: when he became unsettled on the day of his capture he disgorged three young foxes, three young rabbits and three blue-tongue lizards. The first goannas in the Sanctuary in the 1930s were fed a mixture of eggs and milk poured from a teapot through a hose.

Raiders

The setting of the Sanctuary brings both delight – and dangers.

Fleay commented that: 'Situated as the reserve is below the dense timber of Mounts Riddell and Toolebewong, with Coranderrk forest and the Yarra tussock flats to the west and south, there is a vast and continual rallying ground for swift and efficient hunters of the wild, marauding creatures that find the Badger Sanctuary, with its comparatively concentrated and varied population, a veritable mecca.'

Foxes have always been, and still are, a menace to the Sanctuary animals. From the beginning, trapping, shooting and baiting have been a necessary part of preserving the Sanctuary fauna from foxes. They snatch a variety of prey, from young emus to wallabies. The pea-hens and guinea-fowl nesting in secluded places often vanish. The brush-turkeys gradually became fox-wise and built their mounds by the creek sides. Fleay notes in the 1940s that in the winter foxes even scaled the walls of the snake pit and dug tiger snakes out of their burrows, and on one occasion six Cape Barren Geese lay headless in a paddock that was thought to be securely fenced.

Bulldozing wombats often made convenient holes for foxes to enter. A family of foxes was even found within the Sanctuary itself, living in a wombat burrow. Dogs too sometimes found their way in and caused stampedes among the kangaroos. Feral cats stalked the bushland birds and small native animals.

The Australian goshawk hunted pigeons and parrots and young brush turkeys while the brown rat raided the food stores and aviary seed trays. On one rat hunt in the Ingram aviary, Kevin Mason caught 200 by hosing down the holes and hitting the rats as they ran out. Then the rats themselves became food for the owls and eagles.

Sanctuary animals too will make the most of an unexpected feast: 'At least three of the free living greater gliders have come to a violent end during the last few years by misjudging their glides and landing in a Tasmanian devil's open-topped, smooth-walled yard. Needless to say the devils, having no aesthetic viewpoint, left only the tail and fur of their very welcome visitors!'

But it is not just animals that have harmed the Sanctuary stock. From time to time humans have found their entertainment in trying to ride the kangaroos, in stoning the snakes or letting the water out of the platypus pool. Birds and other animals have been grabbed from their enclosures, packed up and peddled for profit. A recent haul of stolen Sanctuary stock, found in a suburban home, included tortoises, a bearded dragon, frogs, two rare parrots and a carpet python with a microchip. A routine scan proved where he really belonged – in the Sanctuary's classroom.

Many of the animals that lived at the Sanctuary in the 1930s and 1940s became well-known characters. Digger, a six-foot grey kangaroo, created his own publicity stunts: 'After one woman had been hugged for the sheer novelty of the sensation, a small bellboy of the Monterey tickled the kangaroo in the ribs. Digger picked the lad up in his arms and hopped off with him until they both fell sprawling over a log.'

It was common at that time for the public to feed the kangaroos and wallabies at the Sanctuary. A movie film of the Sanctuary in Fleay's time shows a kangaroo boxing with a dummy dangled from a wire strung across the enclosure.

Wenda the wombat had games of her own. She used to follow visitors along the bush tracks – until one day she followed some right out through the gate and became the subject of a police search. Her habit of going into the kiosk and pushing the tables over finally curtailed her activities.

The welfare of the animals always had a high priority. The echidna that lived with the lizards 'has a special diet of hen eggs, cod-liver oil, milk and minced meat all mixed together. He does not thrive in captivity, so he is kept only for a little while and then returned to the bush.'

Fleay comments that 'The presence of large aviaries with shows of brilliant, colourful birds are occasionally the subject of criticism but it must be remembered that no government help is afforded this establishment and that the average visitor wants a show for his money and will not walk for miles in the bush searching for wild game.' With only occasional government grants, there was always more need for money.

In the 1930s the Sanctuary increased in popularity year by year, so with government support unavailable, direct assistance was sought from the public, both from organisations and individuals. Funds were donated for enclosures that would have been far beyond the means of the Committee.

The first, planned in 1937 and opened in 1938, was an aviary for eagles subscribed to by readers of the Argus newspaper. Eagles were of great interest to the public because of the controversy surrounding them. Large numbers had been shot because many farmers thought they killed lambs. In reality, rabbits were their usual prey.

Eagles were common in the Healesville district and in Elverd's time the first eagles on show in the Sanctuary lived in a 5-acre enclosure surrounded by a 12-foot fence. The Argus aviary was a bush eyrie 100-foot long and 37-foot high which allowed the eagles ample space for flight. Made of local sawn timber and wire netting, with several dead trees in the enclosure to provide nesting sites, it was thought to be the largest aviary in Australia at that time.

This was soon followed by a new aviary for parrots donated by General Motors–Holden Ltd. 'Measuring 70 ft by 40 ft by 16 ft, the aviary will be equipped with permanent running water, which will cascade into a concrete pool. As far as possible the trees and shrubs have been preserved inside the cage to provide the birds with natural living conditions.' The general public were invited to the opening and 4000 people turned up. GMH provided the tea but it was BYO cup. Company sponsorship has remained a vital ingredient in the Sanctuary's funding of new exhibits.

Healesville residents also sponsored enclosures with the Misses Moreland's cockatoo aviary beside the kiosk and the Lindsay Field Snake Park. Hugh Huxham raised funds for the Sanctuary with theatrical entertainments while other locals organised film nights.

Fleay had many admirers in Melbourne who were keen to see his work continue. Among these were Byron Moore and Edward Green, who both made large contributions to the Sanctuary.

In 1938 Byron Moore provided a new one-acre koala reservation, planted with manna gum and peppermint. In 1939 Edward Green donated funds for a Tropical House, to shelter animals that came from warmer climes than Healesville. The Tropical House was made of cement sheet and corrugated iron and it was heated by water pipes from a coal-fired boiler. The walls and ceilings were packed full of sawdust for insulation. Among its inhabitants were frill-necked lizards, numbats, tree kangaroos, freshwater crocodiles, pythons and Gouldian finches.

One of the inmates was a black-headed python from Townsville. Its preference for snakes as food was discovered when a black snake was taken to the Tropical House for winter warmth and went missing. It reappeared as a corrugated bulge in the python's stomach.

A long-time resident was Pong, a cuscus from New Guinea that was brought back as a youngster by Fleay in 1939. Fleay writes that 'His yellow-skinned face and handsome fur made him an outstanding exhibit. In case there are any doubts as to the suitability of his name, it can be asserted that anyone who has had close acquaintance with an adult male cuscus is never likely to forget it!'

By the end of the decade the Sanctuary had nearly 100,000 visitors and a seemingly bright future – until the second world war took its toll. Petrol rationing stopped cars and buses. Visitor numbers, and therefore the

Sanctuary's income, were reduced to such an extent that there was not enough money to feed the animals. There were constant appeals to the public for snails, worms, grubs, beetles, centipedes and white ants for the lyrebirds and platypuses; and for horses, cows, sheep, vegetables and poultry for emergency feeding.

Some members of the Committee and the community worked hard to support the Sanctuary with endless money-raising schemes from balls and carnivals to gymkhanas and dog shows. Many American service personnel visited the Sanctuary and often contributed generously to its upkeep. Any money there was had to be spent on the animals' welfare rather than maintenance of enclosures. Manpower call-ups robbed the Sanctuary of staff and helpers, so that the general appearance of the Sanctuary deteriorated. There were rumours that the Sanctuary would close – yet it clung on.

Sadly the financial difficulties drove a wedge between Director Fleay and the committee. With the money so tight inevitably there were different priorities. For David Fleay the animals and science came first. He worked at all hours of the day and night and during the war he and his staff took cuts in their salaries so that the animals could be fed.

But the committee was more concerned with the tourist viewpoint and disliked the public criticism of the Sanctuary's run-down appearance and dilapidated fences. Yet as Fleay commented, 'No flesh and blood without money and materials can prevent decay of fences and aviaries.'

Further disagreement arose over the ownership of echidnas Fleay took with him to America in 1947. It was an unfortunate misunderstanding because in his original appointment, under an earlier committee, Fleay had been given complete control over the Sanctuary animals. Apart from those he had brought with him ten years earlier, much of the stock in the Sanctuary had been donated to him or collected by him.

The dispute meant that in 1947, on David Fleay's return from America after a triumphant expedition delivering three platypuses to New York Zoo, he found himself dismissed. The man who had earned a world-wide reputation for Healesville Sanctuary was no longer welcome within its gates.

As the controversy raged in the press, the government was at last stirred into action. It asked Crosbie Morrison to look into the Sanctuary's present condition and its future possibilities.

He found that the stock in March 1948 consisted of 406 individuals of 92 species: 56 bird, 24 mammal, 12 reptile and 10 exotic. This was much less than in pre-war years and Morrison commented that he could no longer say the Sanctuary had the largest and finest collection of Australian fauna. But to reduce costs during the war, some of the animals, including the platypus Rebecca, had to be set free. Without predator-proof fences, many others animals had just vanished.

Morrison criticised some of the enclosures, including the Tropical House, and described the general layout of the grounds as haphazard. Facilities for sick and injured animals were non-existent and he considered the number of staff, only five including the director, totally inadequate.

Administration by a mainly local, honorary committee of management, with no direct access to government, was unsatisfactory. With gate receipts as their only funds, the members were often forced to finance projects themselves or provide bank guarantees.

It was not surprising that Morrison found fault with some aspects of the Sanctuary – it was surprising that he found it there at all.

But his most important finding was that it was of considerable national importance and he recommended that the government should retain it as a State institution and not as an adjunct to Healesville. He felt that it should be developed on much broader lines, with emphasis on research and education; and that the addition of Coranderrk bushland would reduce the 'zoo look' and return it to its original Sanctuary concept.

In 1949, the government, headed by Premier The Hon. Tom Holloway, finally took on responsibility for an institution that was already popular with people from all over the world, and had fostered public interest in native animals.

The new beginning meant fresh blood on the committee, now government appointed with one representative from the Healesville community and one from the Shire Council. Sir Errol Knox, who had been Chairman of Argus & Australasian Ltd when the Argus was the first company to sponsor an aviary, was now to be Chairman of the Sanctuary Committee of Management. The greatest value of the new arrangement was that it guaranteed regular funds.

Jack Pinches, who had been in charge of the Australian section of Melbourne Zoo, had been appointed Director on Fleay's dismissal. But Fleay was reinstated as consultant to the new committee, a position he held until he left Victoria for Queensland in 1952.

After a rough passage, the ark was securely anchored.

THE WORLD OF THE PLATYPUS

*T*oday's platypusary is a far cry from the simple wooden and tin stuctures of the past. The Sidney Myer sponsorship has enabled the platypus to be shown in a quite different way from the first enclosures. A simulated creek behind a sweep of glass, the new platypusary draws the visitor into the platypus's night-time world. In the moonlit ambience, the platypus swims and dives, a dark shape slicing through the water. Under a vaulted ceiling, the curve of display tanks, 40 metres long, combines first-class facilities for the animals with excellent viewing for the visitor.

The creek banks were made from polystyrene, covered with wire mesh, hessian and steel reinforcing rods, and then with dry cement mix. This was carved and painted to create the natural look of a creek bed. Waterfalls tumble over rocks, while ferns, mosses and lichen line the banks.

The different displays house the animals that share the platypus habitat — water-rats, eels and native fish. The sounds of a creek at night are all around you, the calls of frogs and crickets interwoven with the burbling of a stream.

The building, of stringybark timber and mud bricks, nestles alongside the Badger which flows beneath and through it. A giant log supports the entrance: a fallen manna gum given a new lease of life.

Behind the displays there's a huge feeding tank and 30 metres of tunnel system simulating the creek-side burrows platypus dig in the wild. Infrared video equipment allows keepers to monitor the animals' behaviour.

The new platypusary was built at a cost of $1.3 million and took two years to complete.

The Sidney Myer World of the Platypus

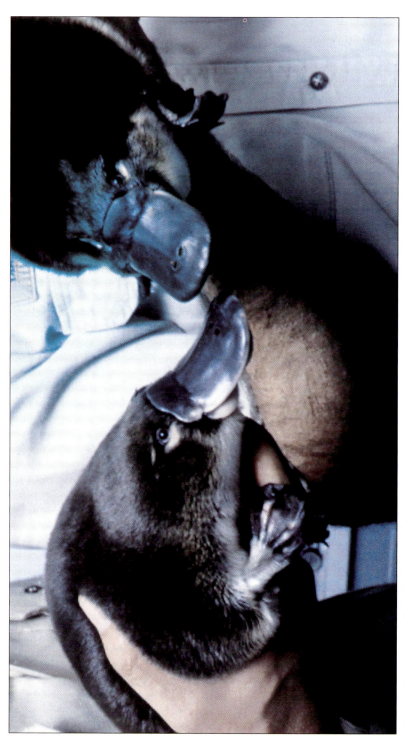

A twin triumph – the first platypus twins bred in captivity

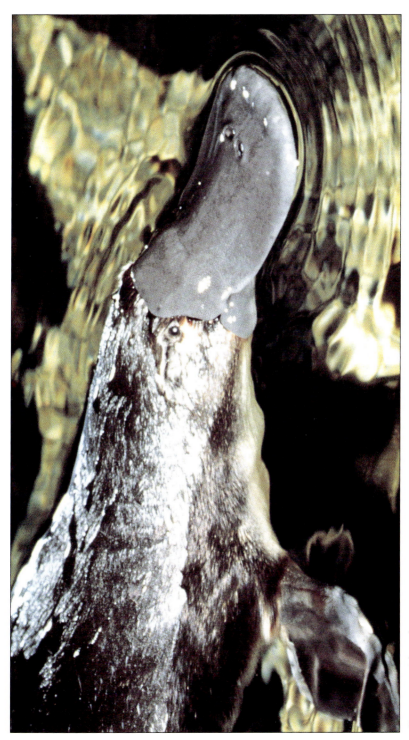

Eye-to-eye with a platypus

Feeding time at the wetlands. They were created by diverting water from Badger Creek, which is also home to the platypus

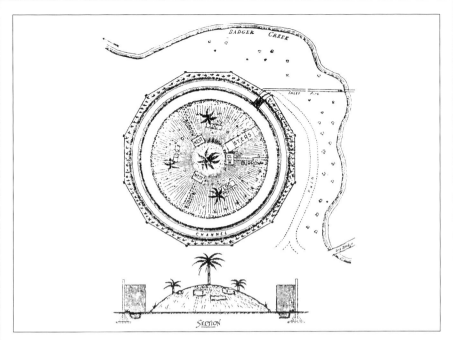

The first platypusary – 1930s

Elverd feeding Glennie with worms

Splash loved to play with a mop

Jill looking for worms

A STAR IS BORN

David Fleay was very keen to breed a platypus, believing this was the only way to get to know the animal's life history. As soon as he could he built 'a primitive platypusary' on a knoll beside Badger Creek.

On a hot February day in 1938, two Healesville men found a tiny platypus, small enough to fit on one's hand, shuffling down the Ben Cairn Road, far away from the nearest stream. Jill, as she was called, settled down well in the Sanctuary, but one afternoon, with a thousand people awaiting her display, she suddenly vanished. She had escaped through a hole in the moisture-rotted flooring planks of her artificial burrow. Fleay found her that night by torchlight, frolicking among the tadpoles of a neighbouring lilypond.

A larger and stronger platypusary, designed to house at least two platypus, was built. Rebecca was full-grown when caught in Boggy Creek and, though she adapted well enough, she was still 'wild' when released in Badger Creek three years later. In the 1939 bushfires, with the neighbouring ranges a mass of flame, Jack was found in a what was left of a pool in Badger Creek. A half-grown youngster, he settled in well and doubled his size over the next year.

Fleay, having provided the best of care, hoped that the platypus would breed. The platypusary, now 90 ft long, included an artificial river bank so that Jill could excavate a breeding burrow. One day in the spring of 1943 Jack seized Jill's tail with his bill and they slowly swam in circles. This courting ritual continued over the next few weeks and on 11 October they were seen to mate.

Great excitement — what would happen next? Did the sounds of excavations mean she was preparing a nursery burrow? On 23 October Jill neglected her food and snatched instead at leaf fragments in the water. Fleay immediately dropped more leaves into the water and these were snatched in her bill. Ducking her head under her body and curling her tail forward, the bundle of leaves was soon held by the tail in a tight grip. Back and forth to the burrow she went, well into the night, carrying load after load of leaves. For the next two days she appeared during the day to feed, and from the evening of the 25th, she vanished for five days.

Could success at last be anticipated? As November dragged on, brief appearances for wetting and grooming, and then longer times for feeding and exercising, kept everyone on tenterhooks. For no-one knew when the baby or babies would leave the burrow for the first time.

By the beginning of January, Fleay was anxious that something may have gone wrong. At the time, it was thought that the young platypus would be crawling about at six weeks, and it was now nine weeks old. So he dug carefully into the nursery burrow – to be greeted by a very angry Jill who, trying to plug up the hole, tossed the nest out, baby and all!

Fleay held in his hand a tiny little platypus, soft-furred and blind with a pudgy tail. But the joy was mixed with anxiety. Would Jill take the baby back? Luckily she did, and the first breeding of the platypus in captivity was a triumph of perseverance and patience.

The news was radioed around the world and made headlines even in London in the middle of the war. The young platypus was a star before she even opened her eyes!

Fleay opened the burrow on several occasions to check her development and at 16 weeks she was shown to the media, now enveloped in long glossy fur, bright-eyed and able to get about. But it was another week before she entered the water and started on the road to independence. When fully weaned at about five months she was nearly as big as her mother.

Christened Corrie, after Coranderrk, she soon became a 'frolicsome, fat and engaging duckbill', pleasing the crowd with her antics and vying with her mother for attention. Fleay adds that 'in the rivalry for possession of food, Jill sometimes clambered upon Corrie's back, pushing her offspring's head firmly below the water in the process!'

Corrie at about 8 weeks, blind, satin-furred and helpless

Fleay's platypusary, 1940s

David Fleay with Jill and Corrie

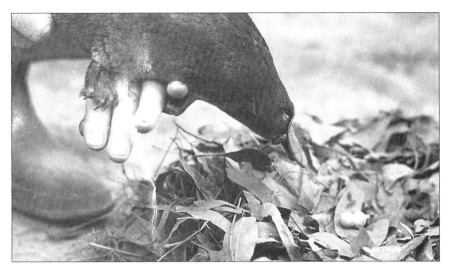

In October 1944 Jill laid two eggs. These were removed a few days later as funds could not be found in wartime to feed any more platypuses

Platypuses from Healesville streams bound for Kangaroo Island in 1940. Holding them are Sigrid Fleay, Cecil Milne and Alf Wright. Sigrid was a great supporter of Fleay's work. Milne and Wright put in long hours to assist Fleay during his decade at the Sanctuary

PLATYPUS TRAVELLERS

Many other platypus from Healesville streams have been housed in the platypusary. In 1940 Fleay was asked to collect platypus for release in the streams of Kangaroo Island, a commission which 'involved a world of darkness in which the water was definitely very cold, our clothes constantly dripping and soft mud clinging and oozing at every step.'

After being caught, the platypus were kept in one half of the platypusary for a few nights to settle them down before their flight, while Jill, Jack and Rebecca occupied the other half. Five pairs, in two lots, were eventually flown to Adelaide and then driven to Kangaroo Island to be released in the streams of Flinders Chase. In their short stay at the Sanctuary they demolished 7000 worms and 250 yabbies. Soon after, a pair were also requested for Western Australia.

These platypus were the first to go by air, but their journey was simple compared to the voyage of a young platypus across the high seas. Winston Churchill had asked John Curtin to send him six platypus in the middle of the Second World War! Fleay persuaded the officials that to send six would be to court a disaster, and it would be better to send one.

A young male was caught and under Fleay's direction a special platypusary, suitable for life at sea, was designed for it. Curtin paid all the expenses and the young platypus — named, of course, Winston — had royal treatment in the months he spent at the Sanctuary in his special home, adapting to his new life. He set sail in September 1943 and made it most of the way. Just four days out from Liverpool, submarines were detected. Sadly, the violent explosions of the HMAS *Port Phillip*'s depth charges killed young Winston instantly, overloading the super-sensitive bill designed to detect a mosquito wriggler on the bottom of a stream.

How did young Winston end up? Stuffed and mounted on his namesake's desk!

DESTINATION — NEW YORK

February 1946 — Please could Fleay travel to New York with a family of platypus. The telegram with this seemingly simple request caught up with Fleay as he combed the wilds of south-west Tasmania on a search for the Tasmanian Tiger.

The New York Zoological Society's passion for a platypus was the start of an enterprise that would engage the energies and efforts of Fleay and his counterparts across the Pacific for the next year.

It began with three weeks' platypus hunting along the rivers and creeks around Healesville. At sunset on cold autumn nights Fleay set his traps along the rivers and creeks round Healesville and then settled down for the long wait amidst the 'snoring chuckles of silver-grey possums, the toy-terrier barks from sugar gliders and sibilant whispers of the ringtails.'

He netted 19 platypus altogether, but only three of these were males, who are normally more wary than females. The animals were kept in the platypusary while Fleay watched and waited to find the animals most suited to undertake the long journey. His final choice was two baby females, Penelope and Betty, and a one-year-old male called Cecil — after Cecil Milne. The other platypuses were returned to local rivers.

Special travelling platypusaries were built for Penelope, Betty and Cecil, but when introduced to their new homes, they lost condition and fretted. Several changes had to be made before the animals settled down, including replacing the galvanised iron swimming pool in the females' platypusary with a 4-inch concrete tank.

It had been intended to take the platypus to New York in June to coincide with warmer weather, but plans in New York were delayed and they spent a year in the Sanctuary. This gave the animals time to become accustomed to appearing before the public, but so much food was required to feed the six now in the Sanctuary that every evening a car had to sent along the Yarra River to collect the 3000–4000 worms dug from the valley silt. Cecil, it was discovered, liked to catch and eat four different species of frog. To prepare for the trip to America, 7000 yabbies and 136,000 worms were frozen in several huge blocks at the Healesville ice works; and 12 large square boxes were built for storing live worms and grubs in soil.

In March 1947, just when all seemed ready for the journey, came a telegram from the Minister of Trade and Customs, Canberra: 'Shipment of three platypus to United States not in the national interest. Export forbidden.' Frantic flying trips to Canberra finally had the order overturned, but the platypus had already missed the ship in Melbourne and now had the extra stress of a plane journey to catch the ship in Brisbane.

The sea journey aboard the MV *Pioneer Glen* was a nightmare for Fleay, although the ship's captain and crew did everything possible to help. The platypus refused all frozen food and Fleay had to radio Pitcairn Island for earthworms. In a midnight rendezvous with the islanders in their whaling boat, worms to refuel the platypuses were hoisted aboard the ship in kerosene tins.

When the *Pioneer Glen* reached the Panama Canal, 10,000 more worms were flown down from New York to Balboa. As the boat was shrouded in fog up the east coast of the States, the blaring of foghorns terrified the platypus. The ship's siren had to be muted to half-strength to pacify the animals.

The ship docked at Boston on Friday 25 April and the platypuses travelled on to New York in a limousine with the two big platypusaries following behind in a truck. But even this part of the trip did not go smoothly. The truck was delayed and the animals had to spend the night in a bath!

When they were finally handed over to New York Zoological Society on Tuesday 28 April, they had as much publicity as any film star. New Yorkers took the duckbills to their hearts and the queue to see them stretched across the Bronx zoo. In the summer of 1947, 4000 people a day flocked to their afternoon display.

Though Betty died in 1948, Cecil and Penelope, fine ambassadors for Australia, were a drawcard at the zoo for the next ten years. The Australian flag and the Stars and Stripes flew side by side at the entrance to the platypus exhibit.

Olympic Tyre Platypusary built in 1955

Platypus swimming in the glass tank

A PLATYPUS LIFE

Since the 1930s, the magnet of the platypus has drawn people to the Sanctuary. In the 1940s, its breeding brought fame and adulation. In the 1950s, it drew sponsorship. And finally in the 1980s, money and technology came together so that we could to begin to unravel some of the secrets of a platypus life.

In 1989, with the help of the Chicago Zoological Society, Dr Melody Serena arrived to study the wild platypus in and around its Sanctuary home. Because of its unique evolution, the platypus was now seen as a World Heritage species.

Designed initially as a 3-year project, Dr Serena's study aimed to look at the social organisation of the platypus and its lifestyle. Do the animals forage over the entire stream, or mostly in deep pools? For how long and when are they active? How many are there and where do they den?

These and lots more questions about the life of the platypus had not yet been answered, for watching them in the dark in a cold creek is no easy task. But by fitting the animals with radio transmitter tags, the movement and activities of unseen platypus could be monitored. Ten animals trapped in Badger Creek were fitted with tags. These were glued to the dense guard fur of the animal's rump just in front of the tail. Signals were sent for up to a couple of months before the tag was shed. A weak steady signal meant the platypus was sleeping in its burrow; in its active periods the signal varied continuously as it swam and dived.

Much was discovered by tracking all the animals day and night. Individuals use numerous burrows scattered along the stream for up to 1.5 kilometres. Their foraging areas overlap and at times two or more animals could be seen calmly swimming within a few metres of each other in the Sanctuary's pelican ponds. Though platypus are mostly active at night, females with young often left their burrows by 4 pm and might not get home till mid-morning.

With the help of Sanctuary guides and FOTZ volunteers, the display platypus were also monitored to discover their patterns of activity. Knowledge of their food intake, the daily weights, air and water temperature, and study of behaviour all help to fill in the picture. To look at their night-life in captivity, a scanning system linked to time-lapse infra-red video cameras was introduced, to detect movement through the tunnels.

Over time, this knowledge will provide a sound scientific basis for understanding and managing platypus populations both in captivity and in the wild.

A platypus takes a dive

A TWIN TRIUMPH

On the eve of the millennium two new stars have appeared – platypus twins, the first twins of their kind ever to be born in captivity. It is 55 years since the only other captive platypus breeding when David Fleay bred Corrie at the Sanctuary in 1943.

Baby X emerged from the burrow for the first time on 3 April 1999. Although breeding behaviour had been recorded, this was the first positive proof of a platypus birth. His brother, Baby Y, took his first view of the outside world six days later. For keepers Norm and Fisk, it was an ecstatic moment, wiping away the many disappointments of years past. For the Healesville Sanctuary it was the culmination of years of work, the triumph once more of persistence and patience.

Mum, Koorina, came to live at the Sanctuary in 1991 when she was about four months old. Her mate, known only as 'N', is also eight years old. They are both locals animals from Badger Creek.

Platypus reach breeding maturity in captivity at about six years old. Koorina and 'N' have shown signs of breeding behaviour since 1994. In the Platypus Breeding Facility housed in the old Platypus display built in 1955, five video cameras are set up to film the animals' day-to-day life. Platypus mating behaviour involves the pair swimming round each other, nuzzling and circling, sometimes with their tails in each other's mouths. This was first observed between Koorina and 'N' on video on 5 November in the 1998 breeding season. Mating behaviour and actual matings continued on 6–8 November. The platypus is generally a solitary animal and 'N' displayed typical platypus behaviour after mating, no longer seeking Koorina's company.

On 14 November, Koorina was seen to gather nesting material, tucking the leaf litter into the curve of her tail. The eggs were laid around 23 November and hatched about ten days later. Koorina did not appear between 23–27 November, but after that ate hungrily, consuming up to one-and-a-half times her bodyweight in food. Baby X and Baby Y will add much to the understanding of their kind and to the pleasure of visitors.

The twins

5

The Olympics 1956 – Victoria on show to the world. The crowds and the cameras descended on Melbourne, and Australian wildlife drew visitors, like a magnet, to the Sanctuary. It was once again a star attraction.

In the years after the government took over there was a gradual improvement in the exhibits and facilities, and the number and variety of animals on display. The Sanctuary was now to be for Australian and New Guinea animals only. Deer and pheasants, rabbits and guinea pigs were removed elsewhere, though a male peacock that escaped capture was allowed to live out his life at the Sanctuary.

But it was the Olympics that inspired the Committee to try to find a new way to display the platypus, an animal already of world-wide interest. The ideal was an exhibit where the platypus could be seen underwater. For such an enterprise the Sanctuary was still in need of sponsorship and this time it came from the Olympic Tyre & Rubber Co. Pty Ltd.

Whereas Fleay had had to design and build his own platypusary, it was now the domain of architects and builders. Various designs were canvassed but that finally chosen was a raised glass tank, the first platypus display anywhere that allowed the public to see the platypus swimming and feeding underwater. The animals had an off-the-scene burrow area to retire to when not on show. A recorded commentary by the well-known naturalist, Crosbie Morrison, explained the fascinating features of a platypus.

The new exhibit attracted great attention and the Sanctuary was soon teeming with tourists. Several thousand Olympic Games participants and officials were invited as guests.

The Sanctuary's popularity with the public grew year by year. Snake demonstrations remained a drawcard and mallee fowl and bustards were added to the birds on display. More staff were employed and several enclosures rebuilt as the early bush timber enclosures had deteriorated. The GMH aviary was in need of renovation and it was now converted to an aviary for wading and waterbirds.

A guide in the 1950s describes the Sanctuary as 'the world's greatest collection of native fauna. Through the protection it affords, many species of our native life have been able to avoid extinction which previously seemed imminent.'

A koala enclosure kept some koalas confined but 'If you look up you should be able to find the two koalas kept in the park among the manna gums. There are enough trees to keep at least two koalas permanently, without the food trees being defoliated.' A 'Koala in this Tree' sign was moved around the Sanctuary to help visitors find them.

There were also a few koalas in the wild state in the Sanctuary. 'They come and go for there are now a satisfying number of koalas along Badger Creek – animals and progeny resulting from breeding for several years in the koala reserves operated at the Sanctuary to re-establish the koala in the district.'

Other animals that had bred at the Sanctuary had been set free. The Victorian Naturalist of December 1955 comments that bandicoots, pademelons and rat-kangaroos are all increasing at a satisfactory rate at the Sanctuary. 'This is very gratifying as so many small marsupials are either extinct or threatened with extinction. Many animals, the progeny of freed animals, are now running wild in the Sanctuary bushland.

The rufous rat-kangaroo (bettong) is a case in point. About 14 inches high, this marsupial is the largest of nine species of rat-kangaroos. It was once abundant along the coast of New South Wales but now its range is greatly restricted. However it is gradually building up its numbers in the wild state in the Sanctuary area. A pair of rufous rat-kangaroos share an enclosure with the long-nosed rat kangaroo (poteroo)...other occupants are three bandicoots, the long-nosed, short-nosed and Tasmanian barred bandicoot. These bandicoots also run wild in the area.'

A walk around the Sanctuary in the 1950s included a suggestion for a stroll along the creek to the picnic tables near Sir Colin MacKenzie's seat, the place he loved to sit on a summer evening. Here a tree-trunk bridge spanned the creek, one of two bridges made from fallen trees. The third footbridge had planks and wooden railings while vehicles had to ford the creek.

By the mid-fifties the numbers of visitors had reached pre-war levels and new facilities were needed. The entrance was moved from its original site, opposite Badger Creek School, to its present position. A new ticket box was built and more car parking provided.

The lattice-work kiosk, which was also a post office in the 1950s, had a new role as a picnic shelter when a modern catering centre was built in 1959. The Brolga Room was added for committee meetings, to entertain VIPs and for all kinds of functions and celebrations. The Van Embdens ran the catering for over twenty years, starting in the old kiosk and gradually seeing an increase in business as the catering centre became established. It was no easy task getting the food ordering right as visitor numbers very much depended on the weather: sometimes it was very quiet and at other times they were run off their feet trying to serve up to a thousand customers in one day.

The Committee of Management held many of their meetings at the Sanctuary, inspecting all the exhibits so that they could see for themselves how things were progressing. Many gave years of devoted honorary service including Major General V. P. H. Stantke, who was Chairman of the Committee from 1950 to 1962. Sir John Jungwirth served for nearly thirty years. He was given the task of forming the first government committee in 1949, was a member himself from that time on and was Chairman from 1962 to 1978.

Despite continual pressure for the Coranderrk bushland to be transferred to the Sanctuary, it was 1955 before this was finally achieved. As most of the Yarra Valley had by now been cleared for settlement or farming, it was considered that this 142 hectares of original forest was most valuable as an area for research and that it should be protected and kept in its natural state. Jack Jones made a major contribution to its conservation, both as Chairman of the Bushland Sub-committee for many years and in detailed surveys of species.

A bush hut built in the 1960s for people working on research projects, has enabled scientists and university students to study many aspects of the plants and animals of Coranderrk. Research into the vegetation has proved the bushland to be of outstanding botanical significance. Its plant species include a wide variety of orchids.

A bushland track through the part of the Sanctuary not taken up by animal exhibits was created in the late 1950s to give people an opportunity to get to know the forest itself. The guide briefly described the natural history, the eucalytps and acacias and some of the more common birds and mammals.

The walk was named Moert Yantha because 'Thus did some southern tribes of Victorian aborigines describe a short walkabout.'

The first Friends of the Sanctuary was started in 1956 with Lady Winifred MacKenzie as President. She continued to support the Sanctuary after Sir Colin's death, establishing the Sir Colin MacKenzie Trust Fund to contribute to 'furthering the scientific knowledge of native fauna maintained in the Sanctuary'.

In 1962 bushfires once more threatened the Sanctuary sweeping across Dalry Road into the Coranderrk and the Sanctuary bushland. The Badger Creek Fire Brigade, with just one Austin fire truck, fought to prevent the fire reaching the Sanctuary animals. Only a last minute wind change saved the day.

A major water-supply system was created by diverting the waters of Badger Creek to supply fire-fighting pipelines and hydrants, as well as the stream through the lyrebird aviary and some more water ponds. These ponds were built to encourage breeding of more free-flying species. The ponds had a varied bird population but a large influx, particularly of Pacific black duck, occurred every year on the day the duck shooting season opened! Three more bridges were built to allow easier access to the different exhibits.

When W. R. Gasking, who had been Director since 1956, handed over the reins to V. C. Mullett in 1963, he pointed out that the Sanctuary was now world-class and Victoria's number one tourist attraction with an annual attendance approaching a quarter of a million.

By the mid-1960s there were 140 species of birds, mammals and reptiles on display in 50 exhibits. The RACV lyrebird aviary was the first walk-through aviary to be planned. During the 1960s this became the trend with a parrot aviary the first to be opened, in 1964. The wader aviary was converted to a walk-through as was the aviary for doves, pigeons, bower birds and quail. Many birds were reported to be nesting. The walk-through aviary which attracted the most visitors was a 'chatterbox of 500 small budgies and 150 small parrots.' A row of aviaries was built to provide breeding facilities for finches and parrots including the rare scarlet-chested and princess parrots.

The Sanctuary still had need for a more constant and reliable supply of water as Badger Creek was often almost dry in times of summer drought. After wide consultation to find the best solution, a large water storage dam was built where Boggy Creek enters the Coranderrk bushland. Completed in 1968, Lake Coranderrk covers about nine hectares and provides a dependable source of water for Sanctuary exhibits – and to fight any future fires.

A Most Desirable Residence

The sacred or white ibis have found the Healesville Sanctuary almost too much of a good thing. Being at home at the Sanctuary, that's great, but when it comes to taking over the whole place, that's a different matter!

In the late 1950s, nestlings from northern Victoria were added to the park's Wader Aviary. From there, twenty-four captive-bred birds were released onto the Sanctuary's pond system. This group attracted wild ibis and by 1963 the wild birds had established a small breeding colony in the trees above the pelican ponds.

Their clamourous squawks and squabbles enlivened the scene, yet by 1978, 700 birds were enjoying life at the Sanctuary and each year the colony was becoming noisier and smellier. The ibis had abandoned their natural feeding habits and had become scavengers instead, raiding the Sanctuary's garbage bins and stealing food from both animals and visitors.

Research showed that the birds weren't just breeding between June and September, they were producing chicks all year round. To add to the problem, when the young became adults they were returning to the Sanctuary to breed.

By 1980 the numbers had reached 1600 and the large stick nests were to be seen everywhere, from the roofs of aviaries to the tops of tree ferns. So, in an attempt to reduce the colony, 150 birds were netted and trucked 150 kilometres away, to the Gippsland Lakes.

They were back in no time, and the ibis had to be persuaded that the Sanctuary was a less desirable place to live. Their food supply was reduced by installing ibis-proof bird feeders and changing feeding schedules so that many of the animals were fed after the ibis had roosted for the night. In the newer exhibits feeders were designed specifically to exclude the ibis: for instance, the termite mound in the echidna exhibit is really a food dispenser, suitable only for the echidna's long sticky tongue.

Activities such as spotlighting to disturb the birds at night, pulling down some of the nests or replacing some of the eggs with plastic hens' eggs, have gradually encouraged many of the birds to make their home elsewhere. The ibis are now once again a welcome, rather than an over-whelming, feature of the Sanctuary.

Ibis nesting on the pelican ponds

One area which had languished in the post-war years was education. In the late 1960s the Lions Club of Healesville donated $3000 for an information centre, and the following year the Sidney Myer Trust $20,000 towards a theatrette for educational activities for both children and adults. The information centre and Theatrette were opened by the Governor General, Sir Paul Hasluck, in 1971. The information centre provided books and displays on natural history for the general public. The first teacher was appointed in 1972 and in the first 6 months a thousand children from fourteen schools made use of the service.

It was time too for the entrance to the Sanctuary to be upgraded. Funds were provided for a new brick entrance, which included a small teacher's room and a souvenir kiosk. It was opened by the Premier, Sir Rupert Hamer, in 1973.

In 1974 the image of the Sanctuary was tarnished by complaints about 'unnecessary deaths'. An enquiry established that many of the deaths were actually orphaned or injured animals, brought in by the public, that were often too young or too badly hurt to be saved. Other deaths were due to foxes, dogs and cats.

Some of the deaths, particularly of the koalas and echidnas, were unexplained, but this drew attention to the fact that there was still much to learn about diseases and parasites in Australian animals.

The report pointed out that facilities for treating sick and injured animals consisted of only one small heated room which was used as a nursery and two outside hospital enclosures for animals receiving treatment. Having an on-site vet was not at the time common practice.

The snake park was unsatisfactory because, though suitable for the summer months, the animals were too exposed in the winter. Once the cold came they hibernated so were not on view to visitors and many of them died. It was not easy to check on the welfare of individual animals.

The difficulties at the Sanctuary were picked up by the press and the undesirable publicity led to a downturn in the number of visitors. Once disenchantment had set in and the perception had arisen that the enclosures were sub-standard and the place in need of a facelift, it took a long time to banish that view from the public's mind. It was also a time when other recreational facilities were being created in Victoria, vying with the Sanctuary for the tourist's time and dollar.

But the criticism also led to the problems being addressed. The Sanctuary was now to come under the Minister of Conservation. The title was changed from Sanctuary to the Sir Colin MacKenzie Fauna Park, since it was considered that the term Sanctuary might be misleading for a reserve where most of the animals were in enclosures.

Funds were provided for a predator-proof fence and for a treatment room for sick and injured animals. A new heated reptile house was already on the drawing board. Completed in 1976, the warm environment of the enclosures meant that the snakes and lizards could be on show all year round. The separate glassed-in exhibits allowed each animal to be looked after individually, and a keeper was appointed who had specialised knowledge of reptiles. Before that, few keepers had had any previous experience with native animals. Up to the time Graeme George took over as Director on Mullett's retirement in 1975, there had been little training for keepers, or feeders as they were called then. George aimed to turn feeders into keepers.

George had first come to the Sanctuary as a youth with the Field Naturalists Club to help build the Nature Trail but had developed his ideas about management of zoos during the last seven years as superintendent of a wildlife sanctuary in the highlands of New Guinea.

At Healesville George dealt directly with the ten keepers who had to keep daily diaries, noting anything of significance in animal behaviour or treatment. A card record system was introduced and every animal had its own card so that they could be individually identified with eartags or birdbands. The local vet now paid a routine weekly visit, starting parasite and disease control programs. He taught the feeders to recognise symptoms, and give routine treatment under his supervision.

More attention was given to the animals' health. The vegetable gardens provided fresh corn and silver beet and other produce for the collection. Oats were grown for the kangaroos and sunflower and millet for bird seed. Dr Pam Whiteley was appointed as a vet, initially part-time, eventually full-time. She was then able to spend more time on preventive care of the animals and to look at the animals' diets as well as treating the sick. Some of the post-mortems could now be done at the Sanctuary instead of sending every animal to the Veterinary Research Institute in Melbourne.

The next challenge was to find a way to show the public the many Australian animals for whom the hours of the night are the times of activity. Fleay's night tours of the Sanctuary were popular in the early days but to be able to share the night world with all the Sanctuary visitors a new approach was needed.

In the 1970s the Victorian government and Philip Morris (Australia) Ltd provided funds for a nocturnal house, a circular underground building, with fifteen enclosures. Here day and night are reversed. By day simulated moonlight tempts many small night creatures out from their lairs to scurry and scamper across their 'forest' floor, or to leap and glide through the trees. At night the lights simulate daytime and the animals sleep.

This has made it possible for visitors to see many of the animals that, because of their nocturnal habit, few people see in the wild. These include bettongs and potoroos, gliders and quolls, and Leadbeater's possums. In recent years, the Victorian Health Promotion Foundation provided funding for attractive new information panels. These emphasise the importance of a healthy diet and describe the animals' natural foods and how these are replicated in captivity.

Breeding enclosures were built to breed mammals for the Nocturnal House, one of the first dedicated off-limit breeding facilities at a zoo. Graeme George was keen to keep breeding populations of these animals to avoid taking any more animals from the wild. He started breeding with squirrel gliders, long-nosed potoroos and brush-tailed bettongs with the aim of creating viable captive populations. George believed that captive populations need to be managed to ensure enough diversity in the gene pool for long-term fitness and survival.

In 1978 the management of the Sanctuary was transferred to the Zoological Board of Victoria as it was felt that the Melbourne Zoo and the Sanctuary had a similar character and purpose and that the Zoo's professional expertise would be useful to the Sanctuary.

The connection to the Zoo opened up new avenues for funding. Money was spent on a stand-by generator and better amenities for the staff. Sewage treatment was extended with reed-bed cleansing ponds to polish the effluent before it got into the creek. With improvements behind the scenes, and a review underway to look at the front-of-house, the Sanctuary was on its way up once more.

In 1983 the name was changed again. Officially it was to be the Sir Colin MacKenzie Zoological Park, but it was to operate under the name that is most familiar to the public – Healesville Sanctuary.

6

TENDING THE ARK

Feeding a family of hundreds, of all shapes and sizes, is like running a first-class hotel with a lot of very demanding guests. The food has to be healthy, but must also satisfy finicky palates. It has to appeal to the animal while providing all the nourishment that the animal would get in the wild. Whether your meal comes in a bucket or a cup-cake frill – well, that depends on your size.

Visit the foodroom at eight o'clock in the morning and the keepers are busy chopping and chipping, measuring and mashing. Each caters for a variety of animals and tastes; from a thumbnail of baby cereal for the pygmy possum to the rabbit carcass for the Tasmanian devil that needs the crunch of bones to satisfy its appetite.

Finding out what animals need to keep them healthy is a continuing process. In the beginning it was inevitably a hit-and-miss affair as little was known of many of the diets in the wild. But over the sixty years the keepers and the vets have built up a wealth of knowledge that ensures the animals have the best possible care.

In the 1930s the local people supplied much of the foodstuff. There were requests in the local paper from councillors and the president of the Healesville Tourist Association for donations. The challenge of feeding the influx of snakes that accompanied David Fleay's arrival at the Sanctuary was taken up with enthusiasm, first by the Sanctuary Committee and later by the public. Each member of the committee was issued with an American-made mousetrap and ordered to immediately search for mice and frogs. 'On Saturday an organised frog hunt was held in the marshes of Coranderrk but the number of frogs captured was disappointing. A standing price of 4/6 a dozen for mice and frogs is offered.'

The public set to with a will and the press reported that 'Fifty dozen mice, birds and frogs have descended on Healesville railway station in response to the appeal for food for the 113 new snakes that arrived a week ago at the Sir Colin MacKenzie Sanctuary. The response has been so overwhelming that officials are concerned now to save the Sanctuary from the food – especially the cost of freight and collection – not to save the snakes from famine.'

Providing for the platypus has always been a problem: the amount of worms needed has taken much money and effort. While trying to breed the platypus in the early 1940s, it was essential to give the animals the very best diet. This meant large quantities of several species of yabbies as well as tadpoles, grubs and earthworms. The cost was about £1 a day in summer when worms went deeper underground. Fleay comments that one young local lad earned enough for a new bicycle and a cow from his worm-digging activities, while an Aboriginal family made enough to buy a new four-wheeled buggy. At one time two men were employed full-time just to dig up worms. When supplies failed 'We just had to dig long and hard by lantern light'.

In Fleay's time 'Meat for a gradually growing collection of quolls, Tasmanian devils, dingoes and birds of prey came from donations of old horses and bull calves, which we shot, skinned and butchered.' Keepers were kept busy shooting rabbits and collecting insects, including moths, crickets and beetles and, the most popular of all with squirrel gliders, witchetty grubs. Fleay comments that, at a time when insects were not readily available, 'We were astounded to find that in spite of ample provision of boiled bread and milk, sweet jam and strips of fatty meat, the squirrel gliders had made a nocturnal attack on a half-grown guinea fowl that lived in their cage.'

For the fluffy gliders, sapwood was collected from the special trees in the Sanctuary that, from the teeth marks in the trunk, were known to be to their liking.

Among the more demanding residents was the numbat that was flown across from Western Australia in 1941 in the hope that more could be learned of its habits. She was offered termites, several species of ants and their eggs, mealworms, beetles, grubs, earthworms, raw egg, bread and milk, honey and jam, but termites proved much the most popular. This hand-sized animal with the 10 cm tongue, lapped up 10,000 termites a day. 'Slowly but surely the old stumps and fallen logs of Badger Creek paddocks and surrounding bushlands were reduced to chips and splinters.'

From the start of the Sanctuary it has taken most of the income at the gate just to feed the animals. When gate takings declined during the second world war, the Herald reported that 'Friends of this distinctive Nature Reserve are rallying to the cause with cash donations and gifts of food…'

Jean Osborne (née Anderson) started work at the Sanctuary in the early 1940s, earning 5 shillings a day. Her daily routine was to feed the cockatoos and galahs, the wonga pigeons and budgerigars; then the birds in the GMH aviary and the egrets and pelicans, goannas and echidnas.

She also looked after the Gouldian finches in the Tropical House. In the feed-shed behind the Tropical House, wheat was soaked in a drum overnight. In the morning, pollen and bran, lucerne and chaff were added.

The fish for the pelicans came in a tin on the train from Melbourne twice a week and the mince came out to the Sanctuary on the mail car. A custard from milk and eggs was made for the platypus and red-tailed black cockatoos. Sometimes emu eggs were used if there were plenty available.

At that time snakes were fed live food. Mice, bantams and guinea pigs were put in the enclosure so that the snakes could hunt when they were hungry. One of Jean's jobs was to spend the afternoons sitting up in a tree outside the snake park to yell at people throwing stones at the snakes, as the snakes were easily damaged.

Keepers were called feeders in the 1960s. But in fact they had to be a Jack or Jill of all trades. A job description might have read 'Someone who is prepared to do anything and everything in any weather. Must be good with animals, good with people, good with plants and a good handyman.' For as well as looking after the animals, the keepers cleaned the toilets, picnic shelters and barbecue areas, propagated and planted new vegetation, and even landscaped the enclosures.

Kevin Mason comments that 'Winters were always a hassle then, with regular frosts that would frequently freeze hoses. Dripping taps would have long icicles hanging from them over the frozen water troughs.' Before he could open up the aviaries, he would have to light matches beneath the locks to thaw them.

His wife Marion was also a keeper: 'Her duties from Monday to Friday were to collect entrance fees from visitors as they walked around, and offer for sale the Sanctuary guide and souvenir books during her daily bird-keeping round. This could become hectic, especially trying to remember who you had sold tickets to and who you had not. The books and a roll of tickets were carried in a leather school bag and the money bag slung over the shoulder. It was no mean feat pushing a small-wheeled garden barrow around with seed and other foodstuffs, contending with emus.' Her wheelbarrow was piled high with lizards and frogs to feed the snakes, mice for the owls, hawks and kookaburras, guinea pigs for the eagles and fish for the ibis and herons. Marion left the Sanctuary to start a family as there was no maternity leave then.

At weekends a casual was employed to sell tickets from a small ticket box, just large enough for one, at the front entrance. It had no electricity, so it was very cold in winter and hot in summer.

Conditions were generally spartan: the staffroom basically a shed with a tin chimney, table and chairs, a meat safe for lunches and a cold-water tap only.

Kevin says there was a high expectation of staff performance. Everyone reported daily to the Director who could hire or fire on the spot. Nearly all the work was done with hand tools, as there was very little equipment. Trenches would be dug with pick and shovel. The only vehicle was an ex-army four-wheel drive and the Director's own lawnmower was used for the Sanctuary's needs.

There were only ten staff, including the Director and two casuals, to look after fifty-four exhibits and over 200,000 visitors a year. The Director was curator, receptionist, public relations, visitor services and first-aid officer, paymaster and vet.

Kevin Mason

Kevin Mason

Kevin Mason's working life at the Sanctuary spans more than half the life of the Sanctuary itself. His first contact with the ways of the wild was as a schoolboy when David Fleay brought a possum or an owl over to the class at Badger Creek. Kevin tells of lunch hours spent in the Sanctuary grounds and afternoons digging up worms for the platypus. He started work as a feeder in the Sanctuary in 1960 at the age of 20.

At that time you were shown round by another feeder for a week and then you were on your own. Everything had to be scrubbed by hand, the paths raked, and the rakes mended until there was nothing left of the original rake. The job included cleaning pavilions and laying fires ready for visitors at the weekends.

One of Kevin's least favourite jobs was tipping the bags of tiger snakes out in the snake pit and then scrambling out of the enclosure as fast as possible. But the job he loved most was caring for the lyrebirds and the new RACV aviary in the early 1960s. He did much of the landscaping of the enclosure as well as tending the plants. Every afternoon was spent in the lyrebird aviary, helping visitors find the lyrebirds in the thick bush habitat, and identifying the wild birds.

Though naturally shy, Kevin was soon enjoying talking to people about the animals in his care. He had had no formal training but, over time, he gained a great deal of knowledge of animals, their behaviour and life histories. His learning was often prompted by questions from visitors, and rounded out by daily experience.

Kevin became expert at handling animals and has answered hundreds of calls from the public ranging from advice about caring for an injured or orphaned animal to fishing a snake out of a backyard pool and rescuing a koala injured by dogs from the top of a tree. He is much in demand by the community for talks to all sorts of organisations and puts in untold hours of voluntary work on behalf of the Sanctuary.

From being a member of the keeping staff, Kevin went on to become Head Keeper and then Special Projects Manager. The Historic Shelter was largely his initiative and he is in charge of the Sanctuary's archives. His responsibilities now cover a wide range, from looking after visitors to liaising with community groups. He assists Friends of the Zoos and the Guide Service in many ways, giving talks and taking groups on night walks. He co-ordinates voluntary work and takes care of staff inductions. The demands on his time come from all quarters – and he is only too willing to help.

Though a devoted family man, he has put the needs of the Sanctuary first all his working life. The animals, the public and the Sanctuary itself have benefited greatly from his dedication.

Though more staff were gradually added there was much less specialisation than nowadays. When Geoff Underwood joined in 1979 he was part of the handyman team of six which looked after maintenance, kept the pathways clear and built drains and barbecues. Later when there was more actual construction work, requiring skilled builders, he transferred to the keeping staff.

Geoff's area encompassed the water-bird aviary, the red kangaroo paddock, the kookaburras, small parrots and doves. And on days when other keepers had days off he also looked after the cockatoos, eagles, large parrots, brush turkeys, hawks and dingoes.

In the first hour all the animals had to be checked and all the enclosures cleaned. Everything had to be ready by 9 am with all the old food picked up and the ponds hosed out. Cleaning was a major part of the job.

The keepers were responsible for everything inside the enclosure and for three feet around. They had to replace gravel on the paths, rat-proof the aviaries with tin around the base, paint and mend holes.

The old foodroom bore little resemblance to the sparkling space that houses the food today. Then the coolroom walls were masonite and often wet so that the walls rotted and sometimes rats broke in. The foodroom keeper made up all the food dishes the night before but the rats often ate them. The rats sheltered in the clay pipes and the firewood stored around the place. The compost was only taken to the worm beds once a week so it attracted flies and rats, ibis and possums.

Geoff Underwood was foodroom keeper in the old foodroom. In the morning he made up the honey water, nectar and so on before going out with the pelican pond keeper to collect branches. 'This used to be done single-handed but then Charlie was up a tree cutting eucalypt branches one day and fell on a powerline. On another occasion he cut his wrist with a bowsaw so now two keepers go together. As well as eucalypts, wattles were cut for gliders and ringtails and 2-foot bark logs for wombats to chew.'

In the afternoon, all the food was prepared for the next day; meat and fish were minced and the seed mixed. 'The animal keeper and the foodroom keeper sorted out the diet between them. Different keepers had different views – it was much more up to the individual keeper then.'

Today's foodroom, sponsored by the Ross Trust in 1983, resembles a state of the art supermarket with hoppers of dog chow and corn, hazel nuts and almonds, trout pellets and seed mixes. The coldstore stocks good quality apples and oranges, carrots and greens. The delicatessen section looks slightly less appetising – unless like many of the animals you fancy a snack of mealworms or fly pupae, or half a mouse!

Flies and crickets are bred in a special room. Fly pupae are a popular snack for many animals while the crickets appeal to reptiles, birds and dasyurids. In another room mice and rats are kept to satisfy the needs of snakes and birds of prey.

These days the diet is carefully tailored to the need and appetite of each individual species. The vets and the keepers work together to ensure that the right nutritional balance is maintained, that the vitamins and minerals the animals need are included every day. A pet food loaf provides a suitable food for many of the animals while the echidna has its own mix which includes minced beef, egg and glucose. There's a special diet sheet for each species.

The koalas still need fresh branches of leaves everyday to keep them active. The keeper looking after them goes out with one of the horticultural team to bring back two-and-a-half truckloads every week. The plantation of 7000 trees in Coranderrk only supplies half their requirements. The rest comes from another 4000-tree plantation in Lilydale or from private properties or trees under power-lines which, for safety, need to be cut down. The koala's need for a constant supply of fresh leaves means that it costs about $5000 to feed a koala for a year.

The old eucalypt branches taken from the koala enclosure have several more uses in the Sanctuary. Kangaroo Island kangaroos and other macropods browse on them, and they provide refuge or camouflage in open paddocks. Cockatoos like to chew on the bark and nuts. The leftovers are chopped up and used as mulch for the Sanctuary's plantings.

It's not just what to feed the animals that's important, but when. Some animals, like the Tasmanian devils, are not fed everyday, to replicate the uncertainty of the hunt in the wild. Snakes will be fed in summer when they are active but not at all in winter when they tend to hibernate.

Some animals have proved exceptionally adaptable in their feeding habits. Ibis have found their long curved beaks to be the ideal tools for helping themselves to everyone else's dinner. Specially designed feed trays have had to be introduced in the hope of outwitting them.

Today the twenty-eight keepers are divided into four teams to cover different sections of the Sanctuary. Training is given in many aspects of animal management, from nutrition to record-keeping, display techniques to communicating with the public. The daily routine of feeding and cleaning and checking every animal is still important. They get to know every animal in their section – in time they can tell the animals apart both by looks and behaviour.

Visitors are discouraged from feeding the animals as the wrong food can harm their teeth, gums or digestion. The animals at the Sanctuary get all the food they need with the right nutritional balance to keep them healthy. The public can be certain that the Sanctuary animals are handled with respect and cared for with dedication.

Male lyrebird on its singing post

A lyrebird on its mound

Male lyrebird displaying

Inside the RACV lyrebird aviary

Helmeted honeyeater with young

Leadbeater's possum

VICTORIA'S ANIMAL EMBLEMS

Two animals have a special place in Victoria — animals which are the State's faunal emblems — Leadbeater's possum and the helmeted honeyeater. Both are endangered, and the Sanctuary is working to conserve them.

LEADBEATER'S POSSUM

Leadbeater's possum was thought to be extinct after not having been sighted for fifty years. But in 1961 it was rediscovered near Marysville and a number of colonies have been found since. Though it was once widespread in Victoria, the possum is vulnerable because only pockets of its specialised habitat remain. Only trees over 190 years old provide suitable nest hollows. A thick under-storey of silver wattles is also necessary to provide wattle sap for food and to allow the possum to move through the forest.

Members of the public have put in many hours of effort searching for the possum colonies — standing in the forest to watch large mountain ash hollows in the hope of seeing a family leaving their home to forage at dusk. Many of the oldest trees, burnt in the 1939 fires, are gradually falling over through age and decay, so it is essential that enough living old trees with hollows are left unlogged. The challenge is to find the balance between the community's need for timber and the Leadbeater's need for a permanent place to live. The Department of Natural Resources and Environment have management plans for the forests that it is hoped will protect enough suitable habitat for the possum.

Because they are so vulnerable, it is essential that colonies of Leadbeater's possums are kept and bred in captivity. The Sanctuary has bred the possum in its breeding enclosures and passed the animals on to other organisations in Australia and overseas, so that there are back-up groups for the wild populations. The Sanctuary's long-term study of the possum is adding to our knowledge of these special animals.

THE HELMETED HONEYEATER

The helmeted honeyeater is the only bird endemic to Victoria. A subspecies of the yellow-tufted honeyeater, it was once distributed in a wide area between Westernport Bay and the Upper Yarra. Its range is now restricted to a small area along the creeks between Cockatoo and Yellingbo to the east of Melbourne.

Numbers have declined from a counted 167 birds in 1967 to a low of 50 birds in 1990. Extinction is a real possibility, and a recovery strategy is in place which involves both the protection of the bird's habitat at Yellingbo and building a captive population at the Sanctuary.

At Yellingbo Reserve the colony of helmeted honeyeaters is monitored and the nests are protected from predators. The work also includes re-establishing suitable vegetation and reducing the numbers of bell miners which compete with the helmeted honeyeaters for territory.

To set up the captive management program at the Sanctuary, helmeted honeyeater eggs and nestlings were initially taken from the Yellingbo population. The first young were fed by hand half-hourly for up to 12 hours a day by devoted Sanctuary staff. Birds of the non-endangered Gippsland race of the yellow-tufted honeyeater are used as foster parents to hatch the eggs and raise the nestlings. Some of the fostered young have become parents themselves.

Despite inevitable setbacks in such a new and complex operation, helmeted honeyeaters are now being bred successfully at the Sanctuary, and the first Sanctuary-bred young have been returned to the wild. Numbers of helmeted honeyeaters have increased to over 100 birds.

The helmeted honeyeater program is a great example of organisations and individuals working together with the common aim of conservation. The State government provided funds and the Department of Natural Resources and Environment looks after the overall management of the Yellingbo Reserve and its helmeted honeyeater population. Universities took on research and genetic and behavioural studies, and inputs came from organisations like the Society for Growing Australian plants and the Bird Observers' Club, as well as local farmers and residents. Friends of the Helmeted Honeyeater, a group of individuals concerned to save the bird, raise funds and assist in habitat restoration and community education.

THE LYREBIRDS

Lyrebirds have always been a major attraction at the Sanctuary. The ranges over-shadowing Healesville are the natural haunt of the lyrebird, and they are quite at home in the Sanctuary's forested enclosures.

In 1939, Edward Green provided the money to build a lyrebird aviary beside Badger Creek. It was an ideal environment: a fully netted creek-side area with its own miniature stream.

The first lyrebirds to inhabit the aviary came for sanctuary, victims of the 1939 bush-fires. Lulu was scorched and tailless. Larrie became remarkably tame and Fleay 'delighted in recording perfect and undoubted reproductions of the nocturnal mopoke call of the boobook owl and the startling shriek of the glider possums in the cock lyre-bird's concert repertoire.'

Though the hoped-for breeding did not eventuate, Lulu and Larrie adopted and reared a baby lyrebird which was later liberated. A lorry driver had found the youngster near Marysville. Within two days Lulu started to feed it, spending all day hunting for worms, grubs and seeds. She stored the food in her beak or throat pouch and approached the youngster with soft grunts. After 14 days Larrie started feeding both the young bird and Lulu.

It is unusual for the male to feed the chick. Usually the mother provides for the young bird until it is about seven or eight months old. In 1985, the female lyrebird at the Sanctuary suddenly died, leaving a ten-week-old chick. To everyone's surprise and relief the father took over the feeding of the chick.

CAPTURING LYREBIRDS

In 1941, David Fleay was asked by the Victorian Game Department to provide lyrebirds for release in Tasmania.

Up in the ranges he found a place where 'the tree ferns closed in, but there were cool and cloister-like passageways between their rough trunks...the whole forest floor was newly scratched over, and at the gully bottom, several narrow passes.'

Snares are made using hazel for springer poles and a soft hemp noose which is drawn round the lyrebird's foot as the hazel springs up. Fleay describes it as 'Finicky and arduous work. Setting twelve snares over a half mile of rough country takes a day.' They captured a female and a few days later a male. The birds were despatched by plane in a tea-chest to be released the same day in a national park 40 miles north-west of Hobart.

Capturing a lyrebird

Packing a lyrebird for the journey to Tasmania

THE RACV LYREBIRD AVIARY

In 1960, when it was suggested to the members of the RACV that they add a shilling to their subscription as a contribution to a new aviary for lyrebirds, the response was overwhelmingly positive. In a matter of months £15,000 was raised, enough to get the project going.

A great deal of care was taken in choosing the site, a secluded area with thick vegetation near the creek, an ideal lyrebird habitat. But fate stepped in. The bushfires of January 1962 swept across the bushland and the very feature that made the site so desirable — its lush clothing of ferns, shrubs and trees — made the fire burn the more fiercely. Nothing was left.

A few months later the structure itself was badly damaged in a storm and had to be rebuilt. A new start was made and at last the huge steel-framed and wire mesh aviary, 62 metres long by 31 metres wide, rose from the ashes. It was tall enough to allow trees to grow to full height and hundreds, some up to 25 ft high, were brought in from nearby forest. Fifty species of trees, shrubs, ferns, creepers and climbers were planted during late winter and spring of 1963. Running water to provide bathing pools and a sprinkler system to maintain humid conditions created a perfect lyrebird habitat — and the most impressive aviary in Australia.

To allow the plants to settle in, the first pair of lyrebirds were not placed in the aviary until November 1964. The spray from the sprinkler system not only promoted vegetation growth, but kept the insect life active in the leaf mulch on the soil surface, allowing the lyrebirds to feed as they would in the wild. Worms from the worm beds were added to the leaf mulch to ensure the lyrebirds had a rich and constant supply of natural food.

Several feeding tables were placed to attract free-flying birds. Finches, wrens, robins and honeyeaters come and go as they please, some even finding ideal nesting sites in the aviary.

In the 1960s, a keeper was present during the afternoons to talk to visitors about lyrebirds and help identify the wild birds and the plants.

LYREBIRD COUNTRY

Kevin Mason has spent many hours in the thick bush which clothes the ranges. He vividly portrays the lyrebird's domain:

> Imagine yourself between the months of May and September beneath the high eucalypt canopy on a mountain slope; rain clouds passing, with moisture dripping from the treetop canopy to the forest floor. As you breathe deeply the aroma of fresh, crisp, earthy air, white clouds of mist drift in the valley below.

> Menura, the male lyrebird, proudly beefs out his territorial call for all to hear, as this is the breeding season. As you draw near you hear a flock of Crimson rosellas, a kookaburra or a yellow-tailed black cockatoo, or perhaps a whipbird — or do you? For it is likely to be Menura, truly a master mimic.

> You may catch a glimpse of a silver shimmer among the lush green ferns as Menura performs a concert. He will perform many times during the day.

In 1963 Kevin was sent out in early summer to trap a pair of lyrebirds for the new RACV aviary. Though he was well versed in bushcraft, he had no field experience, so this was no easy task. First he had to find the birds, looking for their runways and diggings to work out the best places to set the traps. The project involved many days of humping heavy folding wire cages on his back up the steep slopes.

Mirrors were used to entice the birds into the camouflaged cages. The males in particular could not resist taking a closer look at what appeared to be a rival.

A few years later an additional lyrebird was needed for the Edward Green Aviary. On this occasion Kevin's attention was drawn by a young bird's alarm call as it took off through the forest. Kevin chased it along a log and caught it among the ferns. As he put it in a hessian bag, he felt a thump on his head. The female was defending its chick. It attacked again, this time scratching Kevin's head with its long claws, knocking off his beanie. Its third attack was its last — Kevin reached up and grabbed it by the legs. The mother bird too was destined for life in the Sanctuary.

AN INSTANT CELEBRITY

Ric, the male, and Rac, the female (named for the Royal Automobile Club), settled in well and laid their first egg in 1965. That egg was infertile, and the following year the chick died in the egg, but 1967 brought the longed-for event.

Braid (ricrac is a kind of braid) was a healthy chick and was successfully reared, the first lyrebird to be bred in captivity.

Braid was an instant celebrity, and thousands flocked to the Sanctuary in the hope of glimpsing the new arrival.

But Braid was demanding: Can Melbourne people supply enough ants to feed the baby lyrebird?, asked the Press. Melburnians did, and millions of white ants descended on Healesville by all means of transport and in all sorts of containers, even paper bags. Schoolboys brought in ant-infested wood and promised to spend their holidays on the white-ant hunt. Hertz provided a rental car for an 'instant ant pick-up'.

Raising lyrebirds has proved a difficult exercise for Sanctuary staff. As well as the natural hazards of storms and wild winter weather, other animals can be a problem: ringtail possums have discovered that, with small modifications, a lyrebird nest makes an excellent drey while satin bowerbirds find the collection of sticks at the beginning of nest-building a bonanza for their bower construction. Resident pigeons find the top of the nest a great display platform, damaging it in the process. Predatory birds and rodents enjoy the delights of the lyrebird's one and only egg. Even if a young lyrebird is successfully hatched, the chick appears to be very vulnerable in the early weeks of life.

Despite the disappointments in some years, the lyrebirds have bred successfully on several occasions at the Sanctuary. Recently, a new milestone was reached with arrival of Aurora, the first second-generation lyrebird to be born in captivity.

Lyrebird chick

Lyrebird feeding chick at the nest

BIRDS OF PREY

The villains of the piece – that's the way many people view birds of prey. Yet in many ways they do human beings a service: they often eat carrion and most of their prey are animals we regard as pests, such as mice, rats and rabbits.

The Birds of Prey display, sponsored by Australian Eagle Insurance, gives people a chance to appreciate other aspects of the birds' nature, their beauty in flight, their skill in the hunt. But little is seen of the Sanctuary's work behind the scenes or how much time and dedication the keepers contribute so that the birds that can be released have a real chance of survival.

As different species of birds of prey hunt in different ways, understanding the needs of each species is important. The facility for birds of prey often has to accommodate up to 30 birds, from eagles, hawks and falcons to owls. Most are brought in by the public. Some are orphans, some have hit windows, others are the victims of road accidents or shooting. The length of their stay depends on how they are assessed. Some will start training once wounds have healed.

The traditional art of falconry is used in training. Hoods are used to calm nervous birds and leather 'jesses' around the birds' ankles enable their keepers to control them. The bird gradually learns to trust its keeper and to take food from the gloved hand, but before it can be released many weeks may be spent in flying practice. It learns to fly to the lure, and then must prove itself capable of chasing its own prey at the chosen release site. Radio-tracking released birds is gradually increasing knowledge about how rehabilitated birds cope with life in the wild, and the behaviour and movements of each species.

7

Not just the animals

For many people it will be the place itself that lingers longest in the memory. The bushland of Badger Creek is one of the delights of a visit to the Sanctuary. As you wander beside the creek, sit for a while in the fern gully, or picnic in a bushy glade, city tensions ease and peace descends – for as long as you can keep the inquisitive emus and thieving ibis at bay!

Each season has its own feel and fragrance. Spring brings the golden blooms of wattles, a yellow ribbon winding along the creek; the buzz of new life as birds court and nest. Some take to the nest boxes in many of the enclosures but others build to their own design from the dome-shaped grass ball of the superb blue wren, to the tiny wineglass-shaped nest of the rufous fantail. Some families are already out and about, like the brood of maned duck that waddle across the pathway.

In midsummer the delicate white blossom of the Christmas bush still blooms in profusion along the banks of the Badger. The snow-white flowers of the daisy bush mingle with the yellow blooms of the blanket-leaf. The chime of the bellbird rings along the creek but the olive-green birds are elusive, blending into the foliage. From the branch of a blanket-leaf hangs a tiny loosely-woven nest, tied on with cobwebs. The young bellbirds are fed almost constantly by several members of the family.

Autumn is the season when morning mists hug the floor of the Yarra Valley, the hills rising into the clear sky above. The long-fingered light of evening caresses the tree trunks, highlighting the diversity of shapes and shades. A fascinating parade of fungi brighten the forest floor; decaying logs decorated with velvety fans of colour. The satin bower-birds come down from the surrounding hills to feed on the ripening fruit of the lilly-pilly, pittosporum and kangaroo apple.

Winter brings a crispness to the air. A burst of sunlight sparkles on frost-brushed leaves and spangled spiders' webs. An eastern spinebill uses its long curved beak to feed on the nectar of a grevillea bush. Many honeyeaters join the feast awaiting in the flowers of grevilleas and banksias. Birds down from the high country include the currawongs and scarlet robin. The courting lyrebird sings and dances on his mound, the medley of sound floating through the forest. Occasionally snow throws a mantle of white across the landscape.

Kevin Mason has seen it in all its moods and moments:

One aspect that was always special, especially in autumn or after summer rain, was the crisp freshness of the air and the rich fragrance of the bush trees and plants, particularly of the dogwood (Cassina aculeata) and the peppermint trees (Eucalyptus radiata). After heavy rains large blue earthworms would surface and could be seen in large numbers along the pathways…

Memories of magpies and whistling kites nesting in tall gums, with the whistling kites coming to feed from the pelican trough and then alight with fish in claw to the nest of young above. Their frequent calling was part of the atmosphere. Seeing whipbirds, golden whistlers, eastern shriketits, blue wrens and yellow robins in numbers round the kiosk, not to forget the many hundreds of white-naped honeyeaters on a cold winter's morning alight all over oneself or on large logs nearby to feed on honeybread offered…

One could not leave out the flocks of yellow-tailed black cockatoos and gang-gangs; flashes of green and red as large flocks of king parrot, thirty to forty at times, would trustingly feed around the bird tables.

Family groups of kookaburras waited on the bridge railings or large tree stumps to be fed, then chorused their approval, only to be echoed by another group further down.

A ringtail family had its home in the bushes by the long bridge; memories of leaning on the railings watching large trout lazily amuse themselves in the deep pool, with large eels and freshwater crays nearby waiting to be fed some mincemeat, while above, on the still pool waters, the water boatmen skimmed crazily across the surface.

The eels in the Sanctuary Creek began life in the Coral Sea. After drifting down the east coast of Australia, they enter the Yarra River. Travelling by night, they swim upstream, climbing waterfalls and rapids to reach their mountain home. They live in the creek for 10 to 20 years and then swim back to the Coral Sea to spawn.

Watching the water swirling lazily over the rocks, it is hard to imagine that all is not always peaceful along the Badger. But after heavy rain the gentle stream can become a raging torrent, the water clawing at the banks, spilling over, spreading across the landscape, gathering and hurling debris as it goes. But it is forever the Sanctuary's centrepiece, the thread that draws it together and shapes its character.

Sixty years ago the Sanctuary offered:

shadowy tracks, beautiful groves of tree ferns and tall eucalypts which nature planted centuries ago. Off the path from the main entrance gateway one may step into virgin bush tenanted by bell miners, Australian thrushes, robins and many other birds.

Dense scrub walls in the track here and there, with sunny stretches in between, and finger-posts attached to tree-boles lure the visitor from the main 'highway'. In the centre are open spaces with rustic tea-houses, a small museum and seats in the shade of old trees.

It is easy to imagine you are in untamed bushland and not an area enclosed. The illusion is not all illusion for many of these acres have been left in their age-old state and the animals hardly realise that their freedom is curtailed.

Hardly a corner of the zoological park in the hills is not beautiful with trees and shrubs or tall ferns growing wild. Happily little has been done in the way of garden making. The ideal is wildness…

Along the creek manna gums tower above silver wattle and tree ferns, blackwood and Christmas bush. About one hundred and fifty species of plants were known to grow in the Sanctuary in the early days. The twenty species of tree included the dark crinkly-barked peppermint, tea-trees and wild cherry.

But the manna gums which seemed so strong and enduring have proved remarkably fragile. Even by the 1950s, many large old trees in the exhibit areas had to be felled, due to rotting of the main roots and limbs. As the ground has been dug up for pathways or in the building of enclosures, we have caused unseen damage. Root systems have been disturbed, the soil compacted and drainage patterns altered, increasing the spread of disease and the likelihood of wind damage.

Cinnamon fungus has attacked many of the manna gums, causing dieback in the growing tips. Psyllids, sap-sucking insects which live on the leaves, can also stress the trees. The bellbirds that tinkle so melodiously add to the problem by harvesting only the lerps, the insects' little sugary houses, rather than the insects themselves.

The loss of the manna gums affected many other plants that lived beneath its protective canopy, and reduced the living space for the gliders and bats, parrots and owls that made their home in the trees' old hollows.

With the need for removal of unsafe trees and the clearing of areas for animal exhibits and visitor facilities, by the 1970s the Sanctuary had lost a lot of its secluded feel. Free-ranging wallabies and possums had also damaged vegetation. A report in the 1980s described the park as 'looking like a grass paddock with wire cages, bisected by a stream'.

Today wildness is again the ideal. To return the Sanctuary as closely as possible to its bushland best, an ambitious 5-year revegetation project was undertaken, made possible by the generosity of Western Mining Corporation Limited. This sponsorship, the largest in the Sanctuary's history, allowed the employment of expert horticulturalists to head the program.

From the 1950s many plants from other districts had been added to provide a show of native flora full of colour and variety. There were areas specially devoted to displaying native shrubs, to banks of bottlebrushes or mint bushes. In 1955 'some hundreds of native plants not found at Healesville have recently been planted at the Sanctuary'.

The 1957 guide describes exhibit 62 as 'a hedgerow of native shrubs'. The exhibits were numbered up to 94 but many of the numbers in between were left free to allow for future displays.

With long-term funding available the decision was made in 1989 to return the Sanctuary to its original nature and plant only with species that were local natives. The plant communities that once grew naturally within the park have been identified and as they are similar to the communities of Coranderrk, they can be faithfully reproduced.

In Coranderrk bushland there are 365 indigenous species in five main communities so there is plenty of choice for Sanctuary planting. The diverse range of plants will give the Sanctuary habitat long-term vigour and stability and enhance the pleasure of visitors. The revegetation programs are based on the natural succession of plants and the relationship between plants and animals. The environment surrounding the exhibits is as important as the exhibits themselves.

As much of the land on the borders of the Sanctuary had been cleared over the years the first aim was to create a buffer round the Sanctuary and along the new primary pathway. The creek banks too were in need of repair for erosion had worn away the thick vegetation that once lined them. The Fern Gully, developed with the assistance of the Victorian Fern Society, features forty different fern species. It provides a shady resting place beside Badger Creek.

In the first years of revegetation, weeds were a daunting problem but are now much less so as plants have grown and ground covers have become established. Inevitably some species survive more readily than others. Replacing manna gums is a particular problem because they are a favourite food of brushtail possums. When the saplings are about 2 metres tall they can hold the weight of a possum, which often breaks the tree as it strips the leaves.

Seeds are collected from Coranderrk and a nursery has been established in the old vegetable garden. There is a large shade house, poly house, glass house and potting shed, and enough room to hold 40,000 plants. Some plants are contract-grown at a local nursery. The team monitors the growth of seedlings, and is developing the best cultivation techniques in a continuing process that will benefit not just the Sanctuary but similar revegetation programs.

Many groups and individuals have contributed to the Sanctuary's rejuvenation. Year after year in Arbor week hundreds of people, in school groups and families, planted thousands of trees. A trail round the Sanctuary helped students appreciate the intrinsic and aesthetic value of plants and understand their importance to animals. Such a large planting program would not have been possible without helpers like those from the Australian Trust for Conservation Volunteers.

Native tussock grasses provide ideal habitat for native animals such as bandicoots which feed on the seeds and shelter from predators under the tussocks. Since restoring the Sanctuary's lost grasses, many native animals have returned. The native grasses stay lush throughout the summer. A Sanctuary leaflet, Restoring the Balance, highlights some of the Sanctuary plants and the part each plays in the lives of insects and animals.

Sometimes the planting is to provide the right habitat for an animal. Natural vegetation can be used as food in an enclosure. Pasture grasses are trialled for exhibits and holding pens. For the koalas 2 hectares of land in the Coranderrk bushland are planted with swamp gum, manna gum and long-leaf box. The seeds were propagated in the nursery from trees known to be palatable to the koalas.

Many of the 70,000 seedlings planted throughout the park over the last seven years are now well grown and their presence has transformed the Sanctuary. It is once again a Garden of Eden.

Koala with young. The public flocked to see the first koala joeys

The koala enclosure with Director Jack Pinches in the 1950s

Advertising in the 1950s. In the early years Victorian Railways helped to promote the Sanctuary and many visitors came to Healesville by train

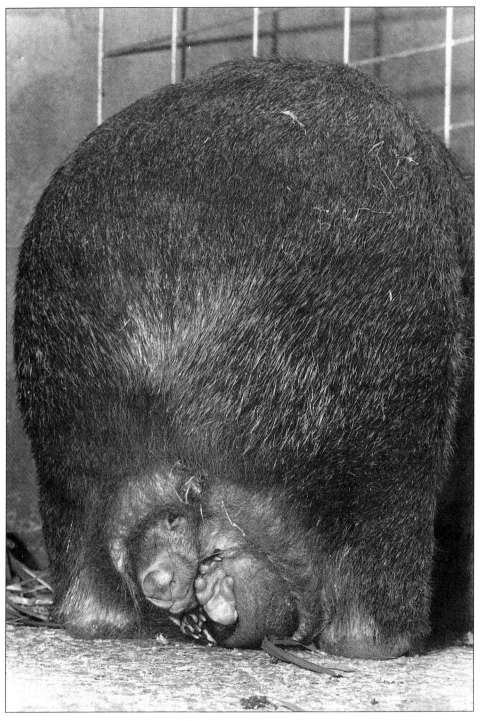

Butch, the first wombat born at the Sanctuary, in 1966

Kevin Mason with Lady Winifred MacKenzie who continued to support the Sanctuary after Sir Colin's death

Graeme George, Sanctuary Director from 1975–1984 with a magpie.

Lake Coranderrk, completed in 1968, ensures the Sanctuary has enough water all year round

The entrance to the Nocturnal House

Mountain pygmy-possum

Long-footed potoroo

Brush-tailed phascogale

Enclosures today aim to provide animals with their natural habitat. The concrete wombat enclosure of the early 1980s contrasts with Wombat Gully constructed in 1988

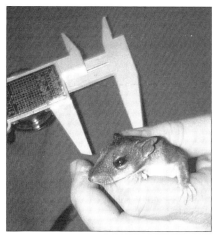

Weighing and measuring animals are a regular part of research activities

Radio-tracking an animal yields valuable information about its lifestyle

The Coranderrk bush hut is a useful base for research activities

Visitors enjoy their interaction with the wombat and the keeper at the Wombat Close-up

Students learning at firsthand about the environment

Artist Syd Tunn with a ringtail possum. Many photographers find inspiration in the Sanctuary animals

Keepers share their understanding of snakes with the public

Hand-rearing orphaned birds is a time-consuming job for Sanctuary staff

This baby echidna was handraised at the Sanctuary

Three orphaned wombats

Vets Dr Rosemary Booth and Dr Peter Holz care for the Sanctuary animals

Wildlife sculptures allow the blind to feel the animals they cannot see

Advertising the Sanctuary in the 1980s

8

THE SANCTUARY TODAY

A shiver of excitement ripples through the crowd as a wedge-tailed eagle swoops low, its great wings almost brushing the visitors' heads. Eyes watch, spellbound, the aerial acrobatics of falcon and hawk, wedge-tailed and sea eagle. A wedge-tailed eagle that has been hand-raised, a peregrine falcon with a missing toe, would be unable to catch their own food in the wild. But they have a new future flying in the Sanctuary – and in the process engendering respect for their fellows with the freedom of the skies.

As the audience marvels at the beauty of flight, the skill of the hunt, visitors learn about the birds brought in shot or injured, about the role of birds of prey in the natural order. 'Where Eagles Fly', the Birds of Prey presentation, is a highlight of a Sanctuary visit. It educates while it entertains, one of the aims and the challenges of the Sanctuary today.

In the reptile house, a python is wrapped around a keeper who handles it with quiet confidence. Some in the audience move forward to touch its cool, scaly-smooth skin. Others look on from a distance, fear and fascination intertwined. The keeper talks about the importance of snakes in keeping rodent populations in balance – and that snakes, left alone, will leave us alone too!

Sharing their knowledge of the animal world with visitors is part of the life of a keeper today. The close contact between the keepers and the public, the animals and the environment is part of what makes the Sanctuary a special place.

The Sanctuary's 50th birthday in 1984 marked a turning point. As at the opening in 1934, an assortment of guests was gathered, welcomed by Dr Alfred Dunbavin Butcher, Chairman of the Zoological Board. As a member of the Sanctuary's Board of Management for the past 35 years, he had seen all the Sanctuary's ebbs and flows of fortune and this Golden Jubilee heralded yet another surge forward.

The government, the shire, the community – all were there to celebrate, congratulate, speculate. Loved as the Sanctuary was, everyone knew that (like most of us after fifty years!) it needed a lift, not just for the face but for the body.

With the government's recognition of the Sanctuary's important role as a tourist drawcard for the State, an extensive list of projects was paraded before the media; and funds were promised to turn dreams into reality. Graham Morris, who had just taken over as Director, believed the Sanctuary needed an overall design to highlight the beauty of its bush environment.

Because the Sanctuary had grown like a jigsaw puzzle – pieces filled in wherever they seemed to fit best at the time – meandering pathways led off in all directions. Though that had a certain charm it also led to a lot of confusion. It was easy to walk a long way and arrive back at the exit having missed the animals you really wanted to see. The time had come for a long-term plan that would give the Sanctuary a sound basis for its future development. The beautiful Badger should be returned to its role as the Sanctuary's centrepiece, with a pathway that would give visitors a chance to enjoy the bush environment to the full. A gravel path now winds its way alongside the creek and log bridges in sweeping curves give the visitor varied views of the water bubbling over the rocks.

The main pathway has all the animals that most visitors to the Sanctuary want to see: koala and kangaroo, platypus and wombat, parrots and dingoes. Smaller secondary loops with a variety of other enclosures provide for the visitor with more time or special interests.

Public attitudes had changed and people were no longer happy to see wombats living in stone and concrete pens. The aim was now to show the animals as far as possible in their own habitat. As a self-help project to raise funds for the new wombat exhibit, a Meet the Animal program was introduced. Visitors were invited to meet wombats, dingoes and eagles as they walked round the Sanctuary. For a small fee, visitors could have their photo taken with the animal. Wombat posters were produced from a painting by Clifton Pugh specially to raise funds. During the six weeks' activity $2000 was raised. When Wombat Gully was opened in 1987, it seemed as though the visitor was sharing the wombat's grassy forest glade.

Animals in the flesh provide a different dimension from even the best wildlife documentary and today a newer exhibit brings the animal so close

Birds of prey presentation

Where eagles fly

Lively graphics at the RACV lyrebird forest

Murrundindi shares his Aboriginal culture through dance

The dingo and its keeper are happy to stop for a chat

The keepers create appetising meals for the animals

Furs and feathers, skins and skeletons – the guides' touch tables attract Sanctuary visitors

that you can feel its coarse thick fur. A young wombat lies on its back snoozing unconcerned as a human youngster strokes its stomach. Wombats seem to take to people, particularly wombats that have been hand-reared. Many young wombats are brought in to the Sanctuary, rescued from the pouch after their mother has been hit by a car.

While the emphasis is on showing the animals to the public in as intimate a way as possible, every care is taken to minimize the stress on the animal. Koalas stress easily, so in the Sanctuary the actual handling is left to the keepers the animals know. But the koalas are close enough for the public to see even the pincer-grip forepaws as one reaches for a fresh leaf, or the 'comb' on the hind paws as another scratches its fur.

You may meet dingoes as you walk around the Sanctuary. They can be well seen in their enclosure but a keeper will often be found exercising one on a leash along a pathway and will be happy to stop for a chat.

Feeding time at the wetlands draws the pelicans and the crowds. The pelicans sail in, jostling for position to be at the head of the queue. As the keeper throws the fish, they catch it expertly in their baggy beaks. Now it's the darter's turn and now the cormorant's. A heron scoops a fish from the water's edge. As the crowd enjoys the lively bird throng, the keeper has a chance to explain the importance of our wetlands and the need to preserve them.

In the new wetlands area, opened in 1989, the birds and the environment are representatives of their kind – the focus of what a wetland would be in the wild. In the natural state, wetlands provide food and shelter for large numbers of our wildlife, from insects and fish to amphibians, mammals and reptiles. In the past 150 years Victoria has lost one third of its wetlands. If the Sanctuary helps people to understand the need for conserving them, it will have gone a long way to fulfilling its role.

Although walk-through exhibits have been part of the Sanctuary for many years, the bush-bird aviary gives a new emphasis to the interdependent relationship between birds and plants, as well as giving us an opportunity to see at close quarters the small birds that many of us have only glimpsed as they flitted by in the bush.

Featuring habitat and its occupants together is the central theme in the newer exhibits: the message is that unless we care for the habitat we will have no birds or animals to care for.

Added enjoyment comes from watching the animals that live freely around the Sanctuary; the finches and honeyeaters that fly in and out of the enclosures to take seed at the feeding tables; the galahs squawking at the picnic ground. An unusual colony is that of the rufous night herons which roost and nest in the wetland area, birds that were released some years ago but find the Sanctuary a comfortable place to live.

But the Sanctuary today is a place of many facets. Though the animal collection is of course the focus, looking after a 1000 animals of 200 species and more than 300,000 visitors involves a complex organisation. It employs over 70 staff from animal experts to accountants, teachers to botanists, scientists to salespeople.

From the bud of a venture which counted its income in pennies, has flowered an enterprise which must raise millions of dollars to fulfil its role. Marketing the Sanctuary so that its attractions are drawn to the attention of interstate and overseas visitors, creating ever more fascinating exhibits so that people from Melbourne will wish to return again and again, helps to ensure that the best can be provided for the animals – and the people.

One of the oldest buildings in the Sanctuary, the first kiosk, is now the Heritage Centre, created in celebration of the Sanctuary's 60th birthday. Here where the early patrons gathered for tea, the Sanctuary story unfolds. As you shelter from a shower or warm up by the log fire on a winter's day, as visitors did sixty years ago, the wall panels around you reflect the ideas and influences that have shaped the Sanctuary over time.

The issues which confront the Sanctuary now are more complex than those which confronted its first supporters. Then, just providing safe places for animals to multiply was thought to be enough to save them from extinction. Now we know that we need much more understanding, both of an animal and its relationship to its habitat, to ensure its long-term survival.

The *Flora and Fauna Guarantee Act 1988* commits the State of Victoria to undertake whatever is necessary to prevent the extinction of wildlife species in this State. Patricia Feilman, Chairman of the Zoological Board of Victoria at that time, believed the Act to be the most important statute to emerge in the long fight to turn back the widespread destruction of fauna and flora over the last 200 years. Nineteen species of animals in Victoria had become extinct during that period and sevety-seven more were on the list of threatened fauna. Feilman commented that zoos are increasingly assuming a stewardship role comparable with the role of other cultural institutions charged with the preservation of natural heritage.

As zoos have taken on the role of guardian so the management of the animal collection has become more complex. With the Australasian Species Management Scheme the details of individual animals are kept on a data base.

This allows for co-operative breeding programs between the zoos so that the populations of rare or endangered species can be managed in such a way as to best protect the diversity of the gene pool. Zoos come together in the best interests of the conservation of the species.

Much thought has gone into deciding which of the many species of animal should be exhibited. The Sanctuary's most distinctive attraction is its environment. To make full use of this exceptional asset it was decided that the Sanctuary should feature the animals that belong in similar surroundings, the animals of temperate south-eastern Australia.

Gary Slater, who, as Curator, helped to develop this policy, felt that it provided the opportunity for the Sanctuary to go more deeply into environmental relationships, 'mine deeper rather than wider'. It will allow the Sanctuary to show animals behaving naturally in their natural habitat. A completely artificial habitat, such as a simulated desert, is no longer on the drawing board.

Caring for the people who come to the Sanctuary is as important as caring for the animals who live there. Human animals also have to be fed. The bistro and takeaway, the shelters and picnic grounds give people the chance to savour the Sanctuary at their own pace and pleasure. Thought is given to the needs of visitors with disabilities so that all can enjoy their day in the bush. Paths are graded for wheelchairs, which can be hired, and animal sculptures around the park allow the visually impaired to experience by touch the shape of animals they cannot see.

The Sanctuary has always belonged to the people. Many individuals and local groups have involved themselves in all aspects of its life, from the pick and shovel work needed in simpler times to the many specialised demands of today. Sponsorships have always been needed to house new animals in the collection, to update enclosures as research pointed to better display techniques or better ways to enrich the lives of the animals. Companies large and small have contributed goods and services. Now individual animal sponsorships allow schools and clubs and families to make a personal contribution to the upkeep of an animal of their choice.

But the contribution is not all in one direction. Some Sanctuary animals have played a useful part in their turn. Peregrine falcons flown at the Ford Motor Company's production plant chased away huge numbers of roosting seagulls which were damaging the paint work on the new vehicles. It took only five days to persuade 90% of the seagulls to roost elsewhere! The cockatoos trialled various materials in their aviary so that Telecom could discover which were suitable for transmitting microwave beams while being resistant to cockatoo attack! The State Elecricity Commission also experimented with overhead cable insulating material that the cockatoos in the wild would not destroy.

Friends of the Zoos – FOTZ – started in 1980 to foster an increasingly wide understanding of all animal life. In 1985 the first group of Sanctuary guides was trained. Now there are 150 guides that volunteer their time to enrich the visitor's experience. They not only take tours but may often be found around the park sharing their knowledge of the animals and the environment with visitors.

The guides look after the Information Centre where furs and feathers, skins and skeletons can be explored. And sometimes they may be found wheeling a trolley of animal artefacts round the park to intrigue or enlighten visitors. They represent the Sanctuary at shows and shopping centre displays. The guides and other volunteers who join Sanctuary Friends take on all sorts of activities from animal watches to helping with holiday programs for families.

Activities arranged for FOTZ members include spotlighting for nocturnal animals and behind-the-scenes tours. The annual kangaroo count in the Coranderrk bushland in November, followed by a barbecue at the bush hut, provides an enjoyable and useful way for FOTZ members to participate. For some, coming and going freely at the Sanctuary, enjoying the environment for an hour or a day, is the greatest pleasure of FOTZ membership.

The Sanctuary is linked to the community at so many different points. At Won Wron Prison Farm in Gippsland aviaries have been constructed where birds of prey can be cared for, under the Sanctuary's supervision, until they are ready to be released. The Sanctuary provides an exciting backdrop for cultural activities and is a rich source of inspiration for artists and photographers. Their work is celebrated in Wildlife Art shows. Aboriginal stories and dance, music floating through the trees at twilight concerts – these are all part of the Sanctuary scene today.

9

For wildlife, a Sanctuary

'The telephone rings loudly, strengthening the shrieks of an outraged cocka-too.

'It just flew straight out in front of me,' says a woman close to tears, as she thrusts a cardboard box on the table. A young man cradling a rosella in his hands is explaining how it 'slammed into the kitchen window', and in the corner a family of four clutch a pillow-case bearing a bewildered baby possum. The phone is eventually answered and through the noise and hubbub a distraught voice can be heard 'But it's in our swimming pool and it's five foot long!'

This description of the Healesville Sanctuary's Animal Reception Centre – on a quiet afternoon – gives some idea of the constant calls on the Sanctuary's expertise. A koala with a broken arm, hit by a car, a pelican caught in a fence, a goanna with a lacerated foot: each day brings new arrivals to the Sanctuary's Animal Reception Centre.

From the early days the Sanctuary has been a haven for hundreds of animals and birds – some unwanted, some injured, others orphaned.

Whether it's a snake in the ducts of the central heating system, a possum stuck in the flue of a wood-heater or a kangaroo entangled in a back paddock fence, the Sanctuary's animal-handling skills are in constant demand. The community can and does turn to the Sanctuary for help whenever a native animal is in need of care. With the back-up of Help for Wildlife, a voluntary organisation devoted to the welfare of native animals, the Sanctuary offers assistance and advice 24 hours a day.

Over the years the Sanctuary has gained a vast store of knowledge. Today the animals that live in the Sanctuary, and those that arrive at its gates in need of help, can receive expert care in the Sanctuary's veterinary hospital, looked after by two vets.

For the inmates of the Sanctuary, the vets' care of the animals extends to all aspects of their well-being: from the food they eat to the space they need, as well as looking after the problems that arise. Much of the work involves keeping the animals healthy, taking blood and faecal samples, checking for parasites. For instance before breeding from the rare orange-bellied parrots it's important to be sure they are free from beak and feather disease.

The one small room which for twenty years served as operating theatre and laboratory, pharmacy and X-ray centre has finally been replaced so that the ever increasing workload can be handled more efficiently. There is always a queue of patients waiting for the vet's attention, a long-footed poteroo needing a biopsy, a dingo with a growth, a swan with sore feet. The swan had spent too much time out of the water chatting to visitors.

Operations are done early in the day so that the animals have time to fully recover while their keepers are with them. It is often the keepers on their early morning rounds who notice that something is amiss: an animal off its food, a limp, a drooping wing. Their practised eyes often pick up the problem early, like the grey teal ducklings with weak legs needing food supplements for rickets or the carpet python suffering from anorexia.

Animals which needed treatment once all had to be caught by hand which often caused a lot of stress. A solution was found with the use of dart guns. David Middleton, the Sanctuary vet in the 1980s, made the first dart guns out of copper pipe with darts made from hypodermic syringes. Dart guns have made it much easier to transport, move and treat the larger animals. The vets today have great expertise in handling animals and their skills are in demand both in the Sanctuary and in assisting other wildlife organisations in the field.

At the Animal Reception Centre most animals are treated, nursed back to health and set free as soon as they are well enough to fend for themselves. They are usually released close to where they were found so they are back in their own territory.

Some animals remain in the Sanctuary if their chances in the wild are diminished. For instance, the albino kookaburra, Albi, has poor eyesight and he would be unlikely to survive for long outside. Birds of prey that are injured may no longer have the ability to hunt but may find a new role in the bird of prey show or be kept for breeding.

Great care is taken not to bring disease into the Sanctuary. Before any animal joins a group in the Sanctuary or is sent off to another zoo, it spends at least 28 days in the quarantine section so that its health can be thoroughly checked.

Many paths lead to the Sanctuary gate. Some native animal pets are abandoned when they get sick or difficult to handle. And some are intentionally injured. Others are brought in after being confiscated by the Customs Service or Wildlife officers. Forestry operations inevitably cause some animals to be displaced, like Milligan, a baby echidna orphaned when a chainsaw cutting through a log killed his mother. At first he was just a pudgy pink blob, the size of a fist, sparsely covered with downy hair. Gradually tiny spines appeared and he learned to lick milk from his keeper's hand, the way he would have licked milk secreted from the glands of his mother's pouch.

The Sanctuary is still a refuge for animals caught in bushfires. Animals injured in the Upper Yarra and Dandenong Ranges fires in 1983 included koalas, echidnas, wombats and ringtail possums.

The most common causes of injury are cat and dog attacks, road accidents and shootings. Road accidents are responsible for most of the orphaned joeys and wombats while possums and gliders are more often the victims of dogs and cats. One baby possum was found clinging to the cat that had killed its mother. The ravages of cats were common even in the early years. Elverd records that he trapped forty cats in the Sanctuary in three years and in Fleay's time: 'A Healesville lady brought me a boot box containing a dozen fluffy ash-grey tails with dark tips. These pathetic reminders of once happy and alert sugar gliders were evidence of a single marauding cat's activities over a short period.'

Despite legal protection, one in ten of the birds of prey brought to the Sanctuary has gunshot wounds. Protected freckled ducks, one of the world's rarest ducks with possibly as few as 8,000 birds left in the wild, are often victims of duck shooters.

Birds make up the greatest number of casualties. They injure themselves by flying into windows, fences and powerlines. Animals get tangled up in rubbish, especially plastics. One wild platypus trapped by plastic debris and another tangled in a fishing line had such severe injuries that nothing could save them. For some animals, too severely injured to be returned to the wild, euthanasia is the kindest option.

With more than 1200 animals brought into the Sanctuary for care each year, the vet is always responding to new challenges. Even when an animal cannot be saved it can often add to the future wellbeing of its fellows by being part of a research project. The Sanctuary responds to many requests from institutions for animal material, and so contributes to the knowledge that will help in the conservation of native animals both in captivity and in the wild.

A Story of Success

For all the frustrations involved in trying to rehabilitate injured wildlife, the story of a badly injured gang-gang cockatoo, brought to the Sanctuary in May 1994, highlights the kind of success that makes it all seem worthwhile.

After first-aid, the bird was left in a hot box overnight in the hope that it would improve. Next day it had further treatment but it was still unable to stand or feed itself, so for the next eight days the staff tube-fed the bird three times a day. It was then decided that the bird would be left to feed itself to see if it had the will to survive. It did have, eating more each day until it was strong enough to be put in a larger outside enclosure while it regained its fitness.

Three days later three wild gang-gang cockatoos flew onto the top of its enclosure calling and making a huge fuss.

Kim White, keeper in the Animal Care Section, wrote:

> It didn't take long to realise what was happening. Our gang-gang was a male while the three visiting birds included a pair and a single female. Obviously they were the bird's social group which had finally found their missing mate.

> We quickly opened the door and released the fully recovered gang-gang. He flew out of the door and on to the top of his old enclosure. The single female gang-gang flew to his side and started to preen his head. A few minutes later they all flew off in synchronised flight back to their home area.

Orphans are either hand-raised or passed on to local wildlife shelters to look after and eventually release. Hand-rearing animals takes a lot of commitment and money. Raising an orphan wombat costs hundreds of dollars just for its special milk formula, Womberoo. A doll's bottle holds the right amount to feed a young wombat, curled up in a pouch of blankets, that was brought in by a family who found its mother dead. The keeper takes the animal home for its night feed, keeping it warm with a hot box.

Hand-raising some animals presents particular problems, like the seven orphaned brush-tailed phascogales or tuans brought to the Sanctuary. How to look after such small creatures which at nine weeks and weighing only the same as a one-dollar coin should have been safely suckling in their mother's pouch?

The problem was solved by fostering them with the Sanctuary's own breeding colony of tuans whose pouch young were older and able to survive outside the pouch. Smeared with the scent of the young of their new foster mothers, the orphans were accepted and soon thriving – the first time carnivorous marsupials had been successfully fostered.

Sometimes it is possible to return young birds to their parents. Co-operating with others working in the wildlife field can bring results. A naturalist identified the site of the nest of a powerful owl chick being treated at the Sanctuary. The parent owls not only welcomed the chick but adopted a second chick whose parents refused to take it back! To achieve such successes, understanding the species is of prime importance. So too is avoiding human imprinting of the young birds while they are being cared for in the Sanctuary. Other young find a new home within the Sanctuary. Boris, a Gould's wattled bat, lived in a Sanctuary classroom after she fell from her mother and landed on a tent in northern Victoria. Unable to fly and totally dependent on milk, Boris was raised by hand. She enjoyed human contact and gave pleasure to thousands of children.

She is not the only animal to have taken an unplanned journey: a green tree frog, Peron's frog, and green tree snake had all arrived at the Sanctuary after being found in boxes of bananas sent down from the north.

Sometimes the public gets over-enthusiastic in its 'rescue' attempts. Every spring dozens of ducklings and other young birds are brought in when mostly their parents would have returned to feed and protect them. In Spring much of the vet's time may be taken up with these new arrivals.

The effort required to treat and rehabilitate injured or orphaned wildlife is often questioned. It is always time-consuming and often unrewarding. But since it is human beings who are so often responsible, whether directly or indirectly, for an injured animal's plight, it is up to us to take responsibility for assisting those that cross our path.

Healesville Sanctuary recognises the intrinsic worth of native wildlife. It believes that its lead will encourage others to share in the task of helping wildlife to survive despite the odds that human habitation has set against it.

10

SPREADING THE WORD

A child holds out a tentative finger towards a carpet python and a shy smile spreads over her face as she feels the cool suppleness of its body and lets go of her fear. Amazement to discover how it 'hears' through its body and senses the children's presence through its flicking tongue.

Children who come to the Sanctuary are inevitably changed by the experience. Going into a Sanctuary classroom is entering a new world. Excitement stirs as they hear a rustle among the leaves, a scurry in the undergrowth.
In the forest setting of the room a furry shape bounds into its nest box among the branches. In the corner a large blue yabby clambers round a tank. Look up – into the gleaming eyes of a wedge-tail eagle, its great wings spread above your head.

From the earliest years, passing on knowledge and stimulating interest in native animals has been the dream of those in charge of the Sanctuary. It's been approached from many different directions, as formal education and as entertainment.

Perhaps because the Sanctuary came into existence at a time when there was recent realisation that a number of Australian animals were under threat of extinction, the desire to foster interest and through interest a desire to conserve native fauna has been a fundamental part of the Sanctuary.

Robert Eadie always took time to share his knowledge with visitors and David Fleay was a natural teacher. The first school groups were taken round in the 1930s. The Sanctuary was the focus of tours for schoolchildren organised by the Victorian Railways Department in 1939 and most of those children had never seen Australian fauna in a natural setting.

Night tours for the general public found an enthusiastic audience.

Greater glider

Night

The crowds have gone and peace descends once more along the Badger. The ibis fly back in from their daytime haunts around the Sanctuary. The chatter of the birds continues until dusk smudges the outline of the trees and darkness enfolds the valley once again.

For a visitor to the Sanctuary 60 years ago:

Over the border of day into the darkness of night, the Sanctuary is enchanted...Stretching out a hand to draw in the extreme overhanging spray of a manna gum, a koala sits in a circle of light from the long ray of the director's torch. One by one they feed, each selecting its own branch, taking perilous chances on the outer tips...

It is a night for lovers. The emu hen drums her mating call. She has dropped her ruff of feathers at her breast and is strutting round and displaying her beauty. The bluff old male answers with a gutteral kind of grumbling note...

All the possum family in the trees and in the yards are out feeding, playing, caring for their young. In the concentrated gleam of the torch we move through the tree shadows from one enclosure to another. A tunnel is cut in the night. The beam catches the bright wide-open eyes of the possums and the beautiful dark shapes of the flying phalangers, revealing their rich colouring and great soft tails. The smooth swift volpane fight is beautiful to watch as the light of the torch follows the possum spread out in the air.

The native cats provide spectacular shows for night visitors, careering round their section of the sanctuary with tails waved high in the air and indulging in all manner of skittish antics; leaping in the air after night-flying insects or climbing trees as actively as possums. Land yabbies burrow up to the surface of the soil at night in all parts of Badger Creek Valley, emerging after dark. Then the native cats reap their harvest, even burrowing down at times to meet the rising yabbies which they extract with their forepaws.

There is much activity in the snake park on a warm night...On the lovely colonnade at the fairy bower by the bridge, billy supper is provided at the end of the tour. And the night birds call and chatter and whistle while the director tells you of the habits and histories of the marvellous animals and birds about you.

Bill Frogley, whose family had a camping park nearby, tells of groups of up to a hundred holiday visitors walking to the Sanctuary by lantern light.

Fleay describes one full moon summer night when 'a most spectacular and entirely unexpected treat was afforded a considerable party of chatting, laughing visitors. Their attention had just been drawn to a long-tailed dusky glider high up against the moon in a lofty gum and no sooner had they seen it than the animal suddenly jumped and swept in a spectacular glide immediately towards us. It gurgled as it flew and following a good 80-yard sweep, landed with a flop on a tree only 30 feet from the onlookers.'

The outdoors was the only classroom then but today the outdoors can be brought inside to give children a more intimate experience. Youngsters gasp with delight as a ringtail possum licks sultanas from their hands; eyes full of wonder as they watch the graceful leaps of a sugar glider – but the smiles fade as they discover how many of these elegant little creatures are torn to pieces by cats. The subject provides a simple introduction to the difficulties faced by so many of our native species. 'And Then There Were None' tackles this subject which is of prime importance because it is the attitudes and actions of the young that will dictate the future of habitats.

Healesville Sanctuary is ideally suited to engage the students in this topic for here the trees under which they will walk, the stream they will cross, are the actual homes of the animals they will see. So the 'No habitat, No homes, No animals' story is highlighted all the more clearly. The Sanctuary too has suffered the ravages of foxes and cats, the cuts of the chainsaw. It has not escaped the damage caused by a mass of humans tramping through the forest, the disturbance to the land caused by the very creation of facilities for animals and people. Here the trade-offs can be easily understood.

The starting point for 'Planet Earth' is that 'We have not inherited the earth from our parents…we have borrowed it from our children.' Children discover that all plants and animals interact with their living and non-living environment and that everything we do affects the planet.

Education *for* the environment not only *about* it, is a central tenet so the encouragement to children and their teachers to go from understanding to action within their local community is inherent. Children are inspired to begin recycling projects to reduce their use of resources, clean up areas in their local environment, mix their own harmless cleaning agents and repellents and make nest boxes for native animals deprived of hollows. As a class they might sponsor an animal at the Sanctuary or join a conservation group.

Activities such as studying the needs and habits of an animal that lived in the local area and then looking at the local gardens to see how well 'your' animal would fare there today gives students a window into the changes that humans make when we move into an area; the harm we do to other animals without even thinking. The very awareness such an activity evokes can be the start of long-term change. And that, in the end, is the Sanctuary's mission.

The teachers are always extending their ideas; finding new ways to inspire the young minds that pass briefly through their orbit. They tie in with and extend any learning opportunities that present themselves.

Each year a special program is devised for Book Week. 'Yakkinn the Swamp Tortoise – the most dangerous year' inspired a program about wetlands. It gave children the opportunity to look at and touch and understand what is special about the long-neck turtle. But beyond that they could explore its typical wetland habitat and take a close look at the tasty water invertebrates that make up its food. Around the turtle and its habitat were woven art and craft activities, as well as drama, giving that connection of all things which is the joy of the bush environment.

Australia's wetlands feature in activity programs in the warmer months and the Frog Bog provides a focus for the study of frogs and their habitat year-round. A program for teachers shows them how students can create a wetland habitat for native wildlife in the school grounds.

In other programs the children play detective: discovering whose fur and bones ended up in an owl's coughed-up pellet, what a lyrebird finds to eat as they imitate its scratchings among the leaf litter. And then there is the 'What ifs' that lead children to realise that all our animal bits and pieces – the way we are fashioned or feathered or furred – make a difference to where we can live, what we can eat, how we can move.

For the youngest there are puppets and puzzles and the exploration of 'What's Up the Creek'. The children look at the range of life in a typical Victorian creek, from dragonflies and insect larvae to platypus, ducks and frogs. The story of Lester and Clyde, the two green tree frogs whose idyllic life is threatened by humans is brought to life when the children are introduced to their namesakes in the classroom amid a chorus of delight. When the Polyglot Puppeteers perform the story of a young tadpole who sets out to rescue her old friend the playtpus who has become ill due to pollution of the creek, another generation of conservationists is born.

Children find a new soundscape in the richness of the Sanctuary's sounds; and can tape them to recall, when back in the city, the bubbling creek, the clamorous ibis in their rookery, dingoes howling, lyrebirds calling and the variety of which each have their own distinctive call.

The Aboriginal connections to the bush and the animals are brought out in many different ways. The children learn that for the Aboriginal people in Victoria 'in the beginning the creator spirit, Bunjil, the wedge-tailed eagle, carved figures from bark and breathed life into them…Bunjil and his brother Mindi, the snake, formed the land that many Aboriginal people refer to as "mother earth".' Murrundindi, elder of the Wurundjeri people whose tribal lands included the Healesville area, shares his culture through dancing, music, artefacts and dreamtime stories.

The stories in 'Dreaming the Indigenous Way' involve animals on display at the Sanctuary. One dreamtime story tells how birds were created when a rainbow shattered into a million pieces. Each small piece of the rainbow had its own special colour, shape and size. And each piece turned into a beautiful bird that was different from all the others. Other stories show the role that the animals played in the survival of the people and the respect people had for the land that nurtured them.

The stories give scope for many activities back at school, from charades based on the stories to understanding the importance of the oral tradition. It gives children a new perspective, a different way to understand. A special program 'Indigenous People – A New Partnership' provides students with an opportunity for greater involvement with Aboriginal people and perspectives.

The Sanctuary has all the drama and beauty needed for every kind of educational fulfilment, from science to the arts: it is the perfect arena for drawing together the practical and the aesthetic, the study of a feather in close-up to the fascination of flight. In looking at the food room and the food requirements, developing food webs, concept maps and graphics, 'Oh! What a Mouthful' combines science and maths, the arts and health. Photography workshops provide subjects of every shape and texture.

A leader in environmental education, the Sanctuary gives teachers opportunities to increase their own understanding of the environment and to develop skills to translate that into all areas of learning. They have a chance to explore at first hand the intricate ecosystem of the Yarra Valley.

Classes cater for about 30,000 children a year and materials and advice help many thousands more to make the most of their visit. For a learning trail for small children, the teacher is provided with a kit bag with books to read, and materials for games, at different spots around the Sanctuary.

The environment has as much to offer as the animals. Here, just beyond the confines of the city, people could be hundreds of kilometres away. 'An

Almost Natural Community' is a program designed for VCE students. It focuses on organisms in their environments, looking at life in an ecosystem and changes in ecosystems. Sixteen hectares of natural bushland allow students to study different communities along the trail, the multi-layered manna gum community along the creek and the narrow-leafed peppermint forest on the drier ground. They can see runs of the swamp-rats in the thick undergrowth of grasses and sedges, look for hollows to suit yellow-bellied gliders, the dreys, or nests, of the ringtails in the thick shrub layer.

Technology too is made use of to spread the word. Conservation conferences with schools have been conducted over the satellite TV network so that children from areas too far away to visit can learn how to monitor local biodiversity and be encouraged to involve themselves with conservation projects in their own areas. The Sanctuary's Education Service has produced a multimedia CD-ROM, Fauna Australis, which features animals and habitats from all round Australia. With colour photographs, animal sounds and maps it's a great resource for people of all ages.

Opportunities to learn are scattered about like seed, to be picked up by those who are interested. As well as the stimulating signage in the exhibits, snappy little facts or riddles are dotted on bridges or trees as mind teasers for the curious. People are encouraged to stop, look, and hopefully enjoy a nest here, a lichen there, the host of diverse life that always surrounds us.

The thread is subtly woven between the animals and the local environment, revealing connections that we all too easily miss, like the sap oozing down the silver wattle's trunk that the sugar gliders love to feed on, or the flowers fashioned for the beaks of honeyeaters. Attention is drawn to the hidden world: the holes chewed out by parrots to make a nest, the inhabitants of the creek lurking in the shadows or curled up in burrows in the bank.

An individual's response to their surroundings cannot be measured only in the data produced, the techniques learned, the knowledge gained. The response to the Sanctuary is as much one of the heart as of the head; to some, like Elizabeth Jackowski, whose poem appears in an anthology created at the sanctuary, it's

> *Where strangers smile happily*
> *as they walk past one another,*
> *delighted with the beauty of nature and her creatures…*
> *Where the prejudices of today's world vanish completely*
> *and the sun seems to shine brighter than anywhere…*

11

Back from the brink

What do you do with a possum that won't breed: stick it in a fridge? Debbie McDonald, the Sanctuary's conservation biologist, tried doing just that. It's one of the ways McDonald tackled the reluctance of the mountain pygmy-possum to add to its kind in captivity. Was it missing the cold temperatures of its snowy mountain home? Would sticking it in a disused drinks fridge – kindly supplied by Schweppes – induce the periods of hibernation it is used to in the wild? Or is the problem the lack of juicy Bogong moths that make up a large part of its diet? Or perhaps the social organisation of the group?

These questions are important because, for vulnerable animals like the mountain pygmy-possum, a viable captive population may be essential to their species' survival. And that means one has to be able to provide the conditions in captivity that will induce them to breed – and that is not always as simple as it seems!

The main threat to the mountain pygmy-possum is the destruction of its habitat for development in the snow fields. As it has such a restricted range, just 2 square kilometres at Mount Hotham in Victoria and 8 square kilometres on Mount Kosciusko in New South Wales, it could easily become endangered or extinct very quickly.

The long-footed potoroo, Victoria's most endangered mammal, was only discovered in East Gippsland in 1978. Reproduction studies undertaken with Monash University, may prove vital for its long-term survival.

Captive populations of endangered animals are usually maintained to ensure a reserve population is available for reintroduction to the wild should some disaster, such as a bushfire, overtake the wild community. Putting some of the animals on display helps to educate the public and enhance awareness of conservation issues. Captive populations also provide researchers with the opportunity to acquire detailed knowledge of each species, knowledge that will be essential to protect the wild populations in the future.

As more species are threatened and habitat destroyed by human activities, it becomes ever more urgent to have a complete understanding of each species' needs. Research is one of the keystones of today's Sanctuary.

As Director Geoff Williams pointed out in 1992, the work involved in research is often long and arduous, sometimes involving all-night studies in inhospitable places under unpleasant weather conditions. But the results of these efforts, he believes, will pay major dividends for the Sanctuary and Australian conservation in the future.

Projects cover a wide field from animal reproduction to nutrition, animal diseases to the techniques for reintroducing animals into the wild. Staff are involved too in habitat research and in making the habitats within the Sanctuary as ideal as possible for the animals living there. University students are often involved in research projects at the Sanctuary, contributing their time and effort and extending their knowledge at the same time.

Infectious diseases in animals are responsible for many deaths and research into their diagnosis and treatment is a continuous part of the work. So too is the refining of anaesthetic techniques for animals that need surgery. Each kind of animal is different in its response. Techniques for collecting semen which can be frozen to preserve as wide a genetic make-up as possible are continually being developed. This will make it possible to preserve genetic diversity without having to take more animals from the wild. Some of the more unusual characteristics of reptiles are being researched for their potential contribution to the development of human medical science.

Because trialling transmitters and other gadgets and techniques is almost impossible in the wild, the Sanctuary animals play an important role, as many trials can take place safely and easily in captivity. Experiments have ranged from testing artificial nest protectors, designed to keep predators at bay, to taste tests aimed at ensuring bait meant for feral cats is not consumed by native animals.

Radio transmitters must be suited to the particular animal and in the Sanctuary situation a harness, implant, glue or collar can be trialled with remote-controlled video, allowing the researcher to watch for any sign of stress or discomfort.

As so many animals brought in to the Sanctuary are later returned to the wild, research continues to monitor the adaptability of animals from the captive to the wild state. When reintroducing animals to the wild, radio tracking is often used to monitor their success. Transmitters attached to the released animals can be used to identify home ranges, nesting and roosting sites and the use of the habitat.

Within the Sanctuary and Coranderrk bushland a number of animals free-range but there is still much to be learnt about the population sizes and the way they interact with their surroundings. Knowing more about the resident bandicoots and wombats, wallabies and emus will ensure that their numbers do not get out of hand.

The rich yet compact area of Coranderrk is an ideal site to study the lifestyles of many different creatures. For instance eleven species of bat have been found there. The bats are caught in harp nets, hitting the harp strings and sliding down into a canvas bag below. As more are caught, tagged and released, the picture of their populations and use of the environment gradually emerges.

In the Coranderrk too there are long-term studies of bushbirds. The CSIRO bird-banding scheme weighs and measures birds and records their plumage and stage of moult, gradually sketching the detail of the birdlife.

Other studies look at the fish life of Badger Creek. Victoria has 200 species of freshwater fish but one third of them are threatened with extinction.

Healesville's situation at the foot of the forest-clad hills of the Great Dividing range makes it an ideal base for learning more about our forest species. All five of Victoria's gliders, from the tiny feathertail to the greater glider, have had a place at the Sanctuary over the years. They delight visitors to the Nocturnal House as they leap from branch to branch with an enviable ease and elegance. The yellow-bellied glider was bred successfully for the first time in 1988. These gliders also occur naturally in the Sanctuary grounds so a project to track the free-range population reveals how much living space they need and identifies where they nest and forage for food. Much field work is carried out on a voluntary basis by Sanctuary staff who often put in many hours beyond their normal working day.

The little sugar glider has bred well in the Sanctuary, and so it was decided to release some into the Coranderrk bushland where there are plenty of nest hollows and abundant food and – so it was thought – no wild population. The gliders were marked with small metal ear tags and reflective tape for identification. Three days later wild gliders unexpectedly appeared and a wild female mated with a released male. The captive-bred animals took to their new life and the knowledge gained from their release will help other animals to be successfully returned to the wild.

Such knowledge is essential if animals are to be restored to areas that they used to inhabit until squeezed out by European habitation. The brush-tailed phascogale, or tuan, is now extinct over more than a third of its former Victorian range. The Sanctuary's captive breeding program makes it possible to produce the numbers to start reintroducing them to their former habitat. But getting it right is not a simple task. Do younger or more mature animals adapt best? Do they need extra food and nest boxes when they are first released? These questions and many others can only be answered by monitoring their progress. Radio collars are attached which allows their movements to be tracked each day.

More than sixty Sanctuary-bred tuans have been released into the forests of East Gippsland where the species has been extinct for several decades. Refining the techniques of reintroducing this species will help the success rate when other species are returned to the wild. Success can often hang on just one thread – predation.

After finding that predators, from foxes and cats to goannas and birds of prey, took a heavy toll of the first animals released, Todd Soderquist, who was working with the Chicago Zoological Society and Healesville Sanctuary on phascogale reintroduction, devised a new method.

At the release site he built cages with holes large enough for just the young phascogales to squeeze through. The Sanctuary mothers, with their newborn litters, were transferred there. When the young were ready to venture out of the nest, they scampered through the forest at night, and as self-taught hunters they seemed to retain the wariness of a wild animal and proved to have a much greater chance of survival.

It is not only forest animals that are in trouble. Since European arrival many grassland species have suffered due to changes in their habitat and predation by foxes and cats, and a number of species have become extinct.

The eastern barred bandicoot almost suffered the same fate, for in 1991 only a handful of animals were left on the mainland, in one tiny colony in Hamilton, Western Victoria. A recovery program involving the community and relevant organisations was set up to save the bandicoot, and the Sanctuary's captive breeding group has contributed to its success. Young raised at the Sanctuary and elsewhere has meant that, as well as protected colonies at Hamilton and Gellibrand Hill, several new wild colonies could be established in the Western District. From under a hundred a few years ago, the population of eastern barred bandicoots now numbers several hundred. The Department of Natural Resources and Environment monitors the release sites. Controlling predators remains an on-going task to enable the bandicoots to survive.

Over the last twenty years there has been much greater co-operation within the zoo community, nationally and internationally, with the aim of protecting threatened species.

The brush-tailed bettong, plentiful along the Murray River last century but now extinct in Victoria, was the first Australian species to be the subject of International Studbook Management. To provide genetic diversity at the start of the program, three animals were brought in from the wild from the small population still surviving in Western Australia.

With the strong emphasis today on protecting endangered species, BHP Community Trust has sponsored a new behind-the-scenes Breeding and Research complex with 26 large breeding enclosures. The latest in technology allows the animals to be videotaped so their behaviour can be monitored by day and night.

Among the animals in the BHP centre is the black-eared miner, Australia's rarest and most endangered bird. The clearing of large areas of mallee for wheat production not only destroyed its habitat but also allowed the common yellow-throated miner to expand its range and hybridize with the black-eared. The first captive-bred clutch of the black-eared miner has hatched and successfully fledged.

Other inhabitants of the Centre are a breeding group of orange-bellied parrots, part of a recovery plan for the species whose numbers are dangerously low. The birds can be seen at the Sanctuary in a special aviary.

Some captive-bred birds have been released into their winter feeding grounds along the Victorian coast. Radio-tracking for nearly two months after release proved that the birds could adapt to their saltmarsh environment. In November orange-bellied parrots take off on the long flight to south-western Tasmania, their summer breeding ground.

Brush-tailed rock-wallabies are another species endangered in Victoria. There are only about fifty left, in two small communities in Gippsland and the Grampians. The expertise of the vets in capturing animals ensured seven wallabies were successfully brought to the Sanctuary. The wallabies will be bred in captivity at Healesville for reintroduction to the wild. Fostering the pouch young with wallabies of another species will enable the rock-wallabies to breed again.

With Melbourne Water's close involvement, endangered swamp skinks were rescued from Tootgarook Swamp and housed at Healesville while the wetland was being restored. Over a hundred skinks were then returned to Toogarook but as they had bred well at the Sanctuary a breeding population is being retained to enable future reintroductions.

Animals are transferred to other institutions to supplement breeding colonies or for research. Blood samples and tissues are provided for external researchers working on a range of projects.

The Sanctuary's expertise is needed in many situations. When an oil spill off the coast of Tasmania threatened to be a disaster for wildlilfe, a Sanctuary team was there to assist in the treatment of 1900 little penguins. Only 60 birds were lost. The Sanctuary now often works with other organisations, with representatives from universities, government departments, other zoos and wildlife organisations and the community in endangered species recovery programs.

Conferences held at the Sanctuary, such as that on Reintroduction Biology, bring people together to share information and experience. In 1996 the Australasian Society of Zookeepers and the Australasian Regional Association of Zoological Parks and Aquaria held their conference at Healesville. Its theme, Zoos – Evolution or Extinction drew people to the Sanctuary from all parts of the world.

There is still much to be learnt, much basic research to be done, to understand the needs of native animals. It seems that it takes the threat of an animal's extinction to trigger action and unlock funds. However, when an animal population is in deep decline the money and effort required to reverse the situation is all the greater. Perhaps in the future we may be wise enough to spend money on prevention; on increasing our knowledge so that we can protect the space and resources each species needs.

If we do that, we may avoid pushing any more animals over the edge. And Sir Colin MacKenzie's desire to make Healesville the world's centre for the study of Australian animals, may yet be fulfilled.

Where to from here?

'One of the future show places of Australia', raved an enthusiastic Sanctuary visitor in the 1930s. This might have seemed unduly optimistic, yet this was the vision of those who fought for the Sanctuary; the inspiration of those who worked for it for no personal gain and often at great expense. And sixty years later the vision has, against all the odds, become a reality.

Is it the Sanctuary that the pioneers envisaged? Not exactly. Morwell Hodges, editor of the *Healesville Guardian*, points out in 1937 that 'the Sanctuary was primarily intended to live up to its name and be a protective reservation for the propagation of certain species of Australian fauna that were rapidly disappearing.'

Though the large area initially envisaged, which might have made a sanctuary in that sense possible, never eventuated, the spirit of wildlife conservation is as strong as ever.

At the beginning the idea was to protect the native animals and birds, to establish reserves for, as Eadie put it, 'the education and enjoyment of the people and to save for those who are to follow us, the world's most interesting animals.' That still holds true.

For three generations, people have come to the Sanctuary for the pleasure of a day out in the bush that young and old can share together; to be in touch with nature and to know more about the other Australians whose destiny is inevitably entwined with our own.

A sanctuary? Yes indeed for the hundreds of sick or injured animals that are brought to its doors each year; for the many endangered species that are being brought back from the brink; some to be returned one day to their home in the wild.

In the frenetic rush of today's urban life, perhaps a sanctuary for people is of equal importance. For one member of the public The Birds of Prey show brings it all together: 'This display is a show stealer and its educational effect marvellous…Bravo, you have given Australian wildlife a true sanctuary and the Australian public a haven to enter and learn the wonderment of our fellow Australians of the gentler variety.'

Healesville Sanctuary's 1994 master plan, shaped by the efforts of all the Sanctuary staff, sets the stage for the future. Drawn together by Director Justin Gamble, it represents a vision – strong enough to endure, flexible enough to bend to the demands of tomorrow.

The vision is clear: 'The Sanctuary is inexorably progressing towards the full realisation of its potential…It is poised to move into the coming millennium…as a stronghold of native ecosystem conservation theory and practice…The rich potential to challenge and positively change the outlook of the visitor is fundamental to the Sanctuary's mission.'

The great renewal of the vegetation over the last decade has brought back the Sanctuary's sense of place. The aim is for the environment itself to be the dominant experience, for the forest to reveal to us the animals that live within it so that it is forever fixed in our minds that the two are inseparable. The animals that share the forest world, or the dim and shadowy world of the creek bed, bog or billabong, will live together in the Sanctuary.

Already coming together, and made possible by a generous bequest, is the Woodland. It is designed to demonstrate the diversity of the plant and animal life that it supports. The scent of eucalypts, the shrilling of cicadas, the laughter of kookaburras – you are steeped in the scents, the sounds, the sights of the woodland world.

Beyond that, Yarra Valley Water is funding the creation of a Mountain Swamp Gum Forest. Here we will come to understand the fragility of its specialised ecosystem: the reliance for survival of plants and animals on the soil, water and each other. The critically endangered helmeted honeyeater, Victoria's bird emblem, will be at the centre of the display. It will be featured among the creatures that share its environment, the frogs and skinks and Leadbeater's possum.

Plans for the future include a walkway through the forest canopy to peep into the life of the treetops; a Night Flight exhibit for owls and bats, moths and gliding mammals; and mini-displays of mini creatures, the world of the witchetty grub or the teeming life in a termite mound. The hidden life of habitats will be revealed: the role of the fungus or earthworm, the crickets or the ants, important players in the scheme of things.

The World Zoo Conservation Strategy (1993), which clarifies the role of zoos in global conservation, points out that caring for our planet's biological systems is one of the greatest challenges to humankind. 'The threats now disrupting natural systems require that zoos make as great a contribution as possible to protecting holistic life systems and to conservation education.' It sees zoos as evolving from the menageries of the 19th century and the zoological parks of today, towards the conservation centres of the 21st century.

It is as a conservation centre that the Sanctuary will fulfil its role: in the way it manages its collection to protect the gene pool of all its threatened species to ensure they have a chance in the future; in its commitment to expanding its knowledge through research, so that captive and free animals live out their lives in a healthy environment; in its ability to lead us to that essential understanding of the natural world, so that we do not destroy it and in the process destroy ourselves.

The Education Service bases its teaching on the philosophy that humans and other animals form part of a natural system and that our attitude to other animals and to the natural world in general will ultimately determine our ability to survive on this planet in a balanced and harmonious fashion.

The *Zoological Parks and Gardens Act 1995* has brought changes in the way the Sanctuary is structured and no doubt changes will continue. But the Sanctuary's commitment to the preservation of biodiversity and the conservation of the environment will remain unchanged. In 1997 the Sanctuary won both the State and national awards for environmental tourism.

In the past many of us had little idea how much we as a community were responsible for the loss of our fauna. Now through the Sanctuary we have the opportunity to know, and to ensure that not another animal is lost. But the conservation of animals and their habitats takes funds and these will only be forthcoming from governments if politicians believe that the community cares enough about the environment to support conservation projects.

As Robert Eadie, asked at the beginning:

Is it too much to hope that after the long welter of animal tragedy there may emerge into the brightness of the future a finer and better understanding between the animals of the wild and ourselves?

If we can do this we may preserve, to some extent, the spirit of the primitive which haunted the plains, mountains and wooded solitudes of Australia, in the mists of the long past and before the dawn of history.

Whatever is going to be done, let it be done now.

Once again it is up to us, the people of Victoria. Animals have no votes.

Albi, the albino kookaburra

Barrett, Charles, 'Australia's "Wild Zoo"', New York Zoological Society *Bulletin*, January–February 1937

Eadie, Robert, *The Life and Habits of the Platypus*, Stillwell & Stephens, Melbourne, 1935

Fleay, David, *Gliders of the Gum Trees*, Bread & Cheese Club, Melbourne, 1947
— *Looking at Animals with David Fleay*, Boolarong Publications, Ascot, Queensland, 1981
— *Nightwatchmen of Bush and Plain*, Jacaranda Press, Brisbane, 1968
— *Paradoxical Platypus*, Jacaranda Press, Brisbane, 1980
— *Talking of Animals*, Jacaranda Press, Brisbane, 1956
— *We Breed the Platypus*, Robertson & Mullens, Melbourne, 1944

Healesville Guardian, 1930s–1960s

Healesville Sanctuary Archives, correspondence, reports, articles

Healesville Sanctuary Guide Books, 1938–1990s

Healesville Sanctuary Masterplan, 1994

Healesville Sanctuary *Tracks*, 1991–1994

Pescott, R.T.M., *The Sir Colin MacKenzie Sanctuary, Chronological History 1920–1978*, 1982 (privately printed)

Proust, A.J., 'Sir Colin MacKenzie and the Institute of Anatomy', *Medical Journal of Australia*, vol.161, 4 July 1994

Sir Colin MacKenzie Zoological Park, Review Group, Healesville, Report, 1983

Sir Colin MacKenzie Zoological Park, Strategy Plan Report, 1985

Thylacinus, vol.19, no.3, 1994

Victorian Naturalist, 1930s–1950s

Wild Life, 1938–1950

Zoological Board of Victoria, Annual Reports, 1978–1997

Zoo News, magazine of the Friends of the Zoos, 1980–1997